ROUTLEDGE LIBRARY EDITIONS:
PLATO

PLATO'S USE OF FALLACY

PLATO'S USE OF FALLACY

A Study of the *Euthydemus* and Some Other Dialogues

ROSAMOND KENT SPRAGUE

Volume 17

Routledge
Taylor & Francis Group
LONDON AND NEW YORK

First published in 1962

This edition first published in 2013
by Routledge
4 Park Square, Milton Park, Abingdon, Oxon OX14 4RN
605 Third Avenue, New York, NY 10017

Routledge is an imprint of the Taylor & Francis Group, an informa business

British Library Cataloguing in Publication Data
A catalogue record for this book is available from the British Library

ISBN: 978-0-415-59194-2 (Set)
ISBN: 978-0-203-10006-6 (ebk) (Set)
ISBN: 978-0-415-62404-6 (hbk) (Volume 17)
ISBN: 978-0-203-10154-4 (ebk) (Volume 17)

PLATO'S USE OF FALLACY

A Study of
the *Euthydemus* and Some
Other Dialogues

ROSAMOND KENT SPRAGUE

Routledge and Kegan Paul
LONDON

First Published 1962
by Routledge & Kegan Paul Ltd
Broadway House, 68–74 *Carter Lane*
*London, E.C.*4

Made and Printed in Great Britain by
Spottiswoode Ballantyne & Co. Ltd
London and Colchester

For

A. L. P.

εἰδότι φωτί

Acknowledgements

MY thanks are due to the American Association of University Women who awarded me the Clara A. Scott Fellowship for the year 1956–1957. As a result I was able to work on the subject of this book at Cambridge University with Dr. A. L. Peck of Christ's College, and at Princeton University with Professor Gregory Vlastos. I am deeply grateful to both of these scholars for their generous gifts of time and sympathy over a period of several years. I wish also to thank Dr. J. M. E. Moravcsik of the University of Michigan, Dr. Hugues Leblanc of Bryn Mawr College, and Dr. Martin Ostwald of Swarthmore College, all of whom have been kind enough to read the appendix, and a number of other scholars who have taken time to talk with me about various aspects of my work, especially Professor Harold Cherniss of the Institute for Advanced Study, Professor W. K. C. Guthrie, Mr. J. R. Bambrough and Mr. R. L. Howland of Cambridge University, Dr. R. S. Bluck of the University of Manchester, and Miss A. N. M. Rich of Swarthmore College.

The English translations of the various passages from Plato and Aristotle are from the Loeb Classical Library, as are the quotations from the Greek, by kind permission of the Harvard University Press and William Heinemann.

Contents

Introduction

THERE is no doubt that there are many fallacious arguments in Plato's dialogues. This book is an attempt to try out the hypothesis that Plato was fully conscious of the fallacious character of at least an important number of these arguments, and that he sometimes made deliberate use of fallacy as an indirect means of setting forth certain of his fundamental philosophical views.

Of course there may well be arguments in the dialogues which Plato regarded as sound but which are in fact fallacious. It is not my purpose to deny this nor to examine passages where Plato may have committed unintentional logical errors. My purpose is rather to insist that in the case of certain specific fallacious arguments Plato was fully aware of the fallacy and used it for a purpose.

As for what is meant by speaking of the way in which Plato 'uses' certain fallacies, the view that I shall develop may be summarized as follows: Plato's general method in these cases is, not to incorporate or build fallacious arguments into his dialogues in such a manner as to produce conclusions based upon unsound foundations, but rather to introduce them for the purpose of working out their implications. Thus he is able (1) to expose them for what they are, and sometimes (2) to clear away possible lines of attack upon his own position, or even (3) to show that when the proper correction is applied his own views receive support.

The book, then, has been called *Plato's* Use *of Fallacy* because it is this latter point, the connection between fallacy and Plato's philosophical position, which has engaged my attention most. But although this subject comes first in importance it cannot come first in order since it is clearly not reasonable to begin to discuss how Plato has used a particular fallacy unless we have previously identified that fallacy. Further, if it is to be maintained that such use was deliberate, the case for consciousness must have been settled in advance.

From the point of view of method, therefore, it seemed to me essential to begin with a dialogue in which a fair number of fallacies could be readily identified, and the controversial problem of consciousness could be quickly settled. The dialogue which satisfies these conditions best is clearly the *Euthydemus.* The two sophists who perpetrate practically all of the numerous fallacies contained in the dialogue are obviously the villains of the piece, whereas Socrates and Ctesippus are equally obviously the heroes. It has not, to my knowledge, ever been suggested that Plato looked on any of these arguments as valid; the dialogue has always been regarded as a piece of scathing satire designed to expose a variety of eristic tricks. What is really surprising is that when similar arguments occur in other dialogues their likeness to arguments in the *Euthydemus* passes unnoticed, or, if it is noticed, the possibility that Plato has made deliberate use of such arguments is not considered seriously.

I therefore begin with a summary and analysis of the entire *Euthydemus.* (I have preferred to deal with the dialogue as a whole rather than to excerpt the passages in which fallacies specifically occur, first, because the dialogue is relatively unfamiliar to students of Plato, and, second, because I believe that these passages are best understood if taken in their proper setting.) Using the *Euthydemus*, then, as a species of control, I have next gone on to discuss important passages in two additional dialogues, the *Theaetetus* and the *Cratylus.* The choice of these passages was dictated by the fact that I wished to concentrate on philosophical points and thus to avoid taking time to argue the case for consciousness afresh in each instance; therefore I decided to choose fallacies whose similarity to fallacies in the *Euthydemus* was close enough to make this unnecessary. In other words, I take it as legitimate to assume that if Plato understands e.g. the fallacious argument against false speaking in the *Euthydemus*, he understands it in the

Cratylus also. The final chapter, on the *Hippias Minor*, is a departure from the general scheme. In this dialogue there occur fallacies which are not precisely similar to arguments in the *Euthydemus*, but these may, I believe, be regarded as conscious from independent internal evidence.

In all three of these chapters the identification of the various fallacies discussed and the decision that their use is conscious is only a preliminary procedure; the major part of each is devoted to an attempt to come to some conclusions as to what may have been Plato's purpose in the employment of these arguments. In each case it will appear that Plato, when he makes use of a fallacious argument, does this in order to call our attention to some fundamental philosophical problem. In this respect he differs radically from the eristic sophist who uses fallacy for the single purpose of achieving victory in the war of words.

The choice of the *Euthydemus* as starting-point has very naturally affected the variety and type of fallacy considered. The sophists in that dialogue are neo-Eleatics, that is, their arguments (which can in the main be reduced to two, equivocation and *secundum quid*) are based upon the philosophical position of Parmenides. As a result the clash between Eleatic monism and Plato's own view is a recurring theme in the first three chapters. The *Hippias Minor*, however, is concerned not with the difficulties in Eleaticism but with some ambiguities in the Socratic ethics; thus a chapter on this dialogue has extended the range of problems which I believe that Plato was concerned to elucidate by deliberate use of fallacy.

The number of dialogues studied is a small one, but as this is very much an experimental essay, it seemed best to keep to passages for which it was possible to make a really strong case for conscious use. (Passages from additional dialogues have, however, been discussed briefly in the conclusion, the appendix, and the notes.) On the other hand, the number of philosophical problems which arise is by no means small: it includes, for instance, the status of particulars, the nature of participation, the distinction between otherness and not-being, the claims of Parmenidean monism and Protagorean relativism to account for learning and memory, the difficulties in resemblance as the basis for a theory of language, and the meaning of ethical terms like 'good' and 'voluntary'. In other words, the study of even a small group of fallacious arguments in Plato can be seen to lead straight to the heart of his philosophy. (Such a study may lead also to interpretations of particular dialogues which are not completely orthodox: for instance, I have become convinced that the *Cratylus* has much more significance for the theory of Forms than is oftentimes supposed.)

I have also added an appendix dealing, in an adversely critical way, with some of Fr. I. M. Bocheński's remarks on Plato's logic. I have done this, not from a spirit of controversy, but because Fr. Bocheński seems to me, in work which is bound to have its effect upon historians of logic, to have underestimated Plato's logical competence very gravely. As far as one can judge, his procedure seems to have been to collect from the dialogues a number of passages containing logical errors and to offer these as evidence on the basis of which we are invited to form an opinion of the extent of Plato's contribution to logic. But although it is useful to have our attention called to passages in which fallacies occur, it is extremely misleading to suggest that an opinion of Plato's logical skill should be based on such passages in isolation: to do so is to place the dialogue, as a literary form, in the same class as the prose treatise. In contrast to Bocheński, therefore, I have usually tried to study Plato's fallacious arguments in the context of whole dialogues, and to give at least some attention to the character of the particular person with whom Socrates happens to be conversing when the fallacy occurs. As the reader will see, the result has been that my estimate of Plato's logical powers is unusually high.

It should also be mentioned that although the *Sophist* is a fine hunting-ground for arguments which also appear in the *Euthydemus*, and although illustrations from this dialogue would have increased the temporal range of my examples, I have not attempted to use the *Sophist* in view of the fact that a number of these parallels have already been noted by Dr. A. L. Peck in his paper 'Plato and the *ΜΕΓΙΣΤΑ ΓΕΝΗ* of the *Sophist*: a Reinterpretation' (*Classical Quarterly*, N.S., vol. II, Nos. 1 and 2, January–April, 1952, pp. 46–7). It was Dr. Peck's paper which first led me to study the *Euthydemus* in detail, and I feel indebted to this article for the notion that some, at least, of the fallacies in Plato's dialogues might be part of a deliberate plan and not the errors of a logical incompetent.

I hope, then, that this book will call attention to the *Euthydemus* as an important but neglected dialogue, and that it will serve as a defense of Plato's understanding of a number of logical points; above all I hope that it will help to recommend the study of fallacy in Plato as a possible approach to the interpretation of his dialogues.

CHAPTER ONE

The *Euthydemus*

FROM the point of view of pure symmetry, the *Euthydemus* is probably Plato's most beautifully constructed dialogue. A brief introductory conversation between Socrates and Crito sets the stage as the dialogue begins. Socrates then relates to Crito a series of five alternating scenes, of which the first, third, and fifth are exhibitions by two sophists, Euthydemus and Dionysodorus, and the second and fourth are exhortations to philosophy by Socrates himself. (The fourth contains an interruption by Crito, which serves to vary the routine.) A final exchange between Socrates and Crito concludes the dialogue.

Introduction, 271A–275D

In the early part of the dialogue we have a number of hints both as to the intrinsic importance of the matter in hand, and to the probable inability of Euthydemus and Dionysodorus to deal with it adequately.

Socrates had been on the point of leaving the palaestra on the previous day when he was prevented by his *daimonion* 272E.* Socrates' spiritual sign does not usually occur in circumstances which are merely trivial, so that it seems unlikely that Plato would represent the *daimonion* as intervening to make the discussion possible unless what followed were to be of real consequence. The reason why Socrates has been detained is speedily made clear by the claim of the sophists to transmit virtue (ἀρετὴν παραδοῦναι) at 273D, since this claim immedi-

* In the case of a single Platonic reference preceded by no mark of punctuation, I have regularly omitted parentheses.

ately raises two of the most crucial Socratic questions: can virtue in fact be taught, and, if so, who are the teachers of it? If eristic had been simply a game, it would not have been worthy of Plato's serious attention, but since its practitioners posed as teachers of virtue, it was a contender for the same activity as his own dialectic, and needed to be exposed in all its shallowness and frivolity. The contrast between the two is subtly implied by Socrates at 274D:[1] he asks Dionysodorus whether the art that sets up to teach virtue (in this case eristic) can also persuade the pupil that virtue is teachable and the persons teaching it (in this case the sophists) are the ones from whom he should learn, or whether some other art be required for this purpose. Dionysodorus replies confidently that no other art is needed, but we can guess that if any art is capable of settling these matters, it will be dialectic, not eristic. This is exactly one of the differences between them, that dialectic contains its own justification,[2] whereas eristic does not and cannot. The sophists will never, by means of their own art, be able to show either that virtue is teachable or that they are its

[1] Strictly speaking, this sentence should read '*Plato represents* Socrates as subtly implying the contrast between the two at 274D' and so forth. In order to avoid circumlocutions of this sort, which would be cumbersome and repetitious, I have spoken throughout of Socrates only. This is, of course, somewhat misleading, but will not, I hope, be misunderstood. It has not been my purpose in this book to distinguish the historical from the Platonic Socrates. My assumption is that whatever happens in the dialogues (with respect to Socrates or any other speaker) was at least *understood* by Plato. Whether or not he originated or even believed the particular point at issue is quite another matter, and one with which I shall not often be concerned. If, then, I speak of Socrates as, e.g., appearing to anticipate that Euthydemus will say so and so, I am not meaning to assert that the historical Socrates need ever have done such a thing or need even have been likely to do it. I am simply employing a short-hand method of referring to something which I feel that Plato depicts Socrates as doing at a particular stage in a particular dialogue.

[2] By saying that dialectic contains its own justification, I mean that Plato appears to regard this art as able, without the aid of any auxiliary art, to bring a person to the point at which he conceives it necessary to choose and follow virtue. We see this in the case of Cleinias, who, when he has for a short time acted as answerer in a conversation with Socrates, is convinced not only that virtue (here wisdom, but see below, p. 12) is teachable 282C, but that it is necessary to pursue it in order to be happy 282D. When, on the other hand, Cleinias acts as answerer for the sophists, he reaches no such conclusions.

teachers; thus their so-called teaching has no basis on which to proceed.

The contrast between dialectic and eristic is, in fact, a recurring theme throughout the entire dialogue. The most obvious and striking form in which this contrast occurs is in the dramatic device of the alternating scenes, but it is evident also in numerous small details. Socrates' interest in the future of Cleinias 275B is a genuine one, for instance, whereas the sophists receive the candidate indifferently; they do not care what happens to Cleinias so long as he is willing to act as answerer for the demonstration. In other words, dialectic (in the person of Socrates) is represented as an art whose practice displays a real concern for the welfare of the individual soul, whereas eristic (in the persons of Euthydemus and Dionysodorus) appears to be completely lacking in any such concern. So long as dialectic and eristic are in conflict over this or comparable points, eristic cannot be dismissed as trivial or as simply a joke. It is in a sense a joke, but a joke with dangerous pretensions.

The first hint that Euthydemus and Dionysodorus will not be able to come to grips with whatever problem may be proposed occurs earlier, in the course of the introductory conversation with Crito. Here Socrates describes the two brothers as having acquired the skill of confuting any argument 'equally well whether it be true or false' (*ὁμοίως ἐάν τε ψεῦδος ἐάν τε ἀληθὲς ᾖ* 272A). In other words, they are not really interested in truth but only in verbal battles (*ἐν τοῖς λόγοις μάχεσθαι* 272A). Socrates' low opinion of the methods of the sophists is also plain from his comments on the fact that they have been able to learn their skill so quickly and at such an advanced age 272C. (The facility with which this kind of questioning is acquired is noted at a number of points in the dialogue, e.g., 300D, 303E, 304A.) Thus Plato's attitude towards the sophists and their facile technique is clear even before he mentions their claim to be teachers of virtue at 273D, and we are well prepared for the failure of this technique to achieve its object before the demonstration begins at 275D.

Scene I 275D–278E

Euthydemus' first question to Cleinias, whether the learners are the wise or the foolish, comes as a slight anti-climax (no doubt designedly so) after Socrates' invocation to the Muses and Memory for assistance in recalling such enormous knowledge 275D. Socrates encourages Cleinias to answer bravely, since perhaps the sophist is doing him the greatest possible service 275E. Dionysodorus, however, whispers to Socrates that the boy will be refuted no matter how he answers.[3] (Here again, Socrates is genuinely concerned with whether Cleinias will benefit from the demonstration, while the sophists are using him simply as a means of displaying their own dexterity in argument.)

In answer to the first question, Cleinias replies that it is the wise who are the learners 275E. Euthydemus, however, points out that when Cleinias was learning from his teachers he did not yet know the things he was learning. Therefore, since he did not know these things, he was foolish, not wise, and his present answer was wrong. Dionysodorus immediately goes on to ask, which boys learnt the piece dictated by the writing-master, the wise or the foolish? (276C) Cleinias answers, the wise, and so contradicts his previous statement. Euthydemus continues with yet another question, which, as Dionysodorus confides to Socrates, is like the others in admitting of no escape: do the learners learn what they know or what they do not know? (276D) Cleinias answers that they learn what they do not know. But does Cleinias know his letters? Yes, of course. Then when anyone dictates, he dictates what Cleinias knows already, since dictation consists of letters. If Cleinias learns anything from dictation, he learns what he knows, not what he does not know, so he is wrong again. But, says Dionysodorus, to learn is to

[3] The reason why Cleinias will be refuted (i.e., sophistically refuted) no matter which answer he gives, is that a question which contains an equivocal term (as we shall see that this question does), is really a form of the *fallacia plurium interrogationum* (e.g., have you stopped beating your wife?). Aristotle is instructive on this point: 'If one does not make two questions into one, the fallacy which depends on equivocation and ambiguity would not exist either, but either refutation or absence of refutation. For what is the difference between asking whether Callias and Themistocles are musical and asking the same question about two people both with the same name? For if one indicates more things than one, one has asked more questions than one. If therefore it is not correct to demand simply to receive one answer to two questions, clearly it is not proper to give a simple answer to any equivocal question, even though the term is true of all the subjects, as some people claim one ought.' (*On Sophistical Refutations*, 175b 39ff., trans. Forster.)

receive the knowledge of that which one learns, and to know is to have knowledge already. Cleinias agrees to this and to the next assertion, that not knowing is not yet having knowledge. And are those who receive anything those who do or do not have it already? Those who do not have it, answers Cleinias, and is refuted again, since Dionysodorus replies that the learners are those who receive knowledge, not those who have it, and are thus the same as those who do not know.

Euthydemus is about to embark on another round of questions when Socrates intervenes 277D. He explains to Cleinias that the sophists are simply preparing him for his initiation by these merry-makings. They want him to find out about the correct use of words (*ὀνομάτων ὀρθότης* 277E) and are pointing out that 'learning' (*τὸ μανθάνειν*) is used both of someone who, having no knowledge of a thing at first, acquires it later, and also of someone who does have knowledge of a thing and uses this knowledge to go on investigating the same thing further. In other words, they want Cleinias to notice that the same word is used for people who are in the opposite states of knowing and not knowing. Socrates calls such questions frivolity (*παιδιά* 278B), since even if one were to learn many or all such tricks one would know nothing more about how matters really stand (*τὰ μὲν πράγματα οὐδὲν ἂν μᾶλλον εἰδείη πῇ ἔχει* 278B). The gentlemen will probably soon give up this sport and get down to the serious business of exhorting Cleinias to the pursuit of wisdom and virtue. First, however, Socrates will give them an example of how he himself would like to see the discussion conducted.

This speech of Socrates (277D–278D) is notable on two counts: it explains the fallacious argument which the sophists have used to refute Cleinias, and it gives us Plato's fundamental objection to such arguments.

All of the refutations in this first scene (refutations which are, of course, not real but apparent) turn upon the fact that a man who learns is in one sense ignorant (that is, he is ignorant of the things he is about to learn) and in another sense wise (that is, he has sufficient intelligence to be able to add the new piece of information to his previous stock of knowledge). Thus Socrates is pointing out that 'learn' is an equivocal term and that the sophists are here employing the fallacy of equivocation.

The contradictions result from the fact that this equivocal term can be used with respect to persons who are in opposite conditions, who are either knowing or not knowing, as Socrates says at 278A. The sophists are clearly using not only the fallacy of equivocation but also another fallacy closely connected with it, the one known traditionally as *a dicto secundum quid ad dictum simpliciter*.[4] This fallacy consists in taking absolutely what should be taken only accidentally, e.g., to go from 'knowing one's letters' to simply 'knowing' (which can also be taken as equivalent to 'wise') or from 'not knowing the things one learns' to simply 'not knowing' (which can also be taken as equivalent to 'ignorant' or 'foolish'). The two fallacies in combination give the sophists a very effective instrument for refutation of the unwary. In the first place they take pains to phrase their initial questions so that Cleinias will have to choose between a pair of exhaustive alternatives. The equivocal nature of 'learn' then enables them to produce the opposite of whichever one he chooses. If he says the learners are the wise, or those who know, they bring forward a case in which those who learn are ignorant of some one particular thing. They then drop this qualification and assert that the learners are those who are ignorant as such, or those who do not know. If on the other hand he says that the learners are the foolish, the procedure is reversed: the sophists bring forward some particular thing the learner does know and then go on to say that because he knows this thing, he is knowing or wise. Of course

[4] The fallacy is described by Aristotle (*On Sophistical Refutations* 166b 38ff., trans. Forster):

> Fallacies connected with the use of some particular expression absolutely or in a certain respect and not in its proper sense, occur when that which is predicated in part only is taken as though it was predicated absolutely (*ὅταν τὸ ἐν μέρει λεγόμενον ὡς ἁπλῶς εἰρημένον ληφθῇ*). For example, 'If that-which-is-not is an object of opinion, then that-which-is-not is'; for it is not the same thing 'to be something' and 'to be' absolutely. Or again, 'That-which-is is not, if it is not one of the things which are, e.g. if it is not a man'. For it is not the same thing 'not to be something' and 'not to be' absolutely; but owing to the similarity of the language, 'to be something' appears to differ only a little from 'to be', and 'not to be something' from 'not to be'.

Aristotle then goes on to describe the converse fallacy, *a dicto simpliciter ad dictum secundum quid*, e.g., an Indian is black but he has white teeth. Therefore it is (wrongly) argued that, because of his teeth, he is both white and not white. (In the *Euthydemus* there is an example of this fallacy at 299Bff.)

in saying that the learners are the wise, Cleinias presumably thought himself to be asserting that people who are able to learn are intelligent, not that they are omniscient; thus he was not rightly refuted by admitting there to be some special matter which the learners do not know. But since his remarks appeared to have led to the conclusion that the learners are the ignorant, and since wise and ignorant are clearly opposite, he believed himself to have been refuted when such was not in fact the case. Again, when he said the learners were foolish, he of course meant that they were ignorant of what they were about to learn, not that they were utterly stupid, so that again his refutation was not real but apparent.[5]

[5] Mr. Richard Robinson, in an article, 'Plato's Consciousness of Fallacy', *Mind*, LI, No. 102, April, 1942, pp. 97–114, has given some attention to this passage, and comes to the (to me, at least) extraordinary conclusion that although Plato quite obviously regarded the *Euthydemus* arguments as fallacious (p. 102) this discussion of the equivocal nature of *μανθάνειν* gives us no reason to believe that Plato realized 'the peculiar subtlety and formidableness of this type of fallacy'; it contains nothing more than 'one of those crass ambiguities out of which puns are made' (p. 107). In fact, 'Plato appears to have remained till death at the point of view stated in the *Euthydemus*, that ambiguity [i.e. equivocation] is of no importance to the philosopher' (p. 114—the concluding sentence of the article). As far as I can see, Robinson's chief reason for asserting that Plato attached no importance to the fallacy of equivocation is that there exists neither in the *Euthydemus* nor elsewhere in the dialogues any technical term (I suppose *ἀμφιβολία* or *ὁμωνυμία*) for this fallacy. If there is no technical term for equivocation, nor, in fact, for fallacy in general, Plato can have had no consciousness of this fallacy nor of fallacy as such (pp. 102–3). And where there is no abstract consciousness of a thing, this thing cannot, apparently, have been thought important. (See his *Plato's Earlier Dialectic*, Oxford, 1953 (2nd ed.) pp. 4–5, for a general exposition of this view.) I would submit that since Plato is perhaps, of all philosophers, the least addicted to the use of technical terminology, the absence of such terminology proves nothing at all. One might as well argue that Plato attached no importance to metaphysics, since he has no name for this part of philosophy. In an earlier article in *Mind*, L, No. 98, April, 1941, pp. 140–155 on 'Ambiguity', Robinson also refers to this passage in the *Euthydemus*. Again he writes that Plato shows consciousness of ambiguity here, but goes on to say that at 278B such matters are dismissed as puerile. 'This profound contempt for the jugglers of language, and this serene confidence that their art can have nothing to do with the serious business of science, are dominant in Plato's early dialogues' (p. 141). Plato's contempt for the jugglers of language is certainly profound, and it is certain that he refers to their tricks as *παιδιά*, but he does this, not because he thinks that their art has nothing to do with 'the serious business

As we have seen, the verdict of Socrates on this sort of activity is that it is frivolity, and is so for the reason that it tells us nothing about the real state of things. This is a just accusation, since the sophists, in order to place Cleinias in the position of appearing to contradict himself at every turn, have been forced to change their own ground each time as well. Their contribution to the question they have raised concerning the relation between learning and wisdom (if we are charitable enough to suppose that they have really raised such a question) is exactly zero. They have used words as counters in a game, not as correct indices to things; hence they know nothing of the *ὀνομάτων ὀρθότης* which Socrates ironically pretends that they are demonstrating. We may guess, too, that if virtue is thought to be knowledge, any art which produces no knowledge will obviously fail to lead to any kind of virtue—in fact it is not, properly speaking, an art at all. The art which does convey knowledge and which leads at least to the choice of virtue, is given a demonstration by Socrates in the next scene—a demonstration which could hardly contrast more sharply with the one which has just occurred.

Scene II 278E–282D

Do all men wish to prosper (*εὖ πράττειν*) begins Socrates, with an apology for asking such a simple-minded question. This is obviously what all men wish, but how will they bring it about? Clearly by having many good things, such as wealth, health, physical beauty, noble birth, talent, and honor. In addition, men will wish to be temperate, just, and brave, and to possess wisdom. This seems to be a complete list of the goods men need in order to prosper, but one has been left out, namely, good fortune, which is really the same as wisdom. That these are in fact the same is illustrated by various examples to which Cleinias agrees: flute-players have the best luck in flute-playing, school-masters in writing, pilots in navigation, and so

of science' but because (1) the consciousness of ambiguity which the sophists undoubtedly possess could be put to serious use, but in fact is not, and (2) their arguments stand in the way of the development of his own positive philosophy. That Plato conceived this fallacy and others as having a great deal to do with 'the serious business of science' is one of the major contentions of this book.

on. In every case, wisdom causes men to be fortunate (she can never make a mistake and still be wisdom) so that he who has wisdom has no need of good fortune as well.

A man will be happy and prosper if he possesses many goods, continues Socrates 280B, but only if he receives some benefit from these goods. A thing cannot benefit us unless we use it, so that our happiness will depend on the use of our possessions. Furthermore, it will depend on their *right* use, since it can be shown that more evil results when a man uses a thing wrongly than when he leaves it strictly alone. As far as right use is concerned, nothing seems to bring this about except knowledge. It seems, then, that we cannot get any benefit from our other possessions unless we have this one. If a person lacks knowledge, it will be better to have and do little than to have and do much, since much action with no sense can only result in evil and misery. The whole question of goods, then, is not whether they are good in themselves but whether they are guided by wisdom or by ignorance. Other things are not in themselves good or bad, but wisdom is good and ignorance is bad.

The conclusion must therefore be 282A that we should endeavor in every way to become wise, since wisdom is the means to the right use of things, and it is their right use that makes us happy. Cleinias agrees that he ought to be willing to perform any honorable service in order to become wise. Of course, says Socrates, wisdom may not be teachable, but may present itself to men of its own accord. For his part, Cleinias thinks it is teachable, thus saving Socrates a long inquiry into this point. If Cleinias thinks both that wisdom can be taught and that it is the only thing in the world that makes men happy, he surely cannot help saying that it must be pursued, and, in fact, intending to pursue it with all his might. This, says Cleinias, is exactly what he intends to do.

In this second scene it is clear from the start that Socrates will be talking about 'how matters stand', not simply playing with words, since he follows up his first assertion, that all men wish to prosper, with the question, *how* are they to prosper. His discourse will lead Cleinias to the choice of something (the means to prosperity), whereas the demonstration by the sophists led to nothing but his verbal confutation. After this beginning, Socrates has only to show what he thinks the means to prosperity is (he thinks, of course, that it is wisdom), and Cleinias will not only *say* that it should be chosen, but will actually be disposed to choose it; right speech will lead to right action.

Socrates' first question is of especial interest in connection with right speech since it contains the equivocal expression εὖ πράττειν, a phrase which can mean either to fare well (to succeed) or to do well (to act rightly). However, Socrates makes no attempt to play on the ambiguity of the expression, as the sophists did in the case of 'learn' in the previous scene. It is of course more than likely that Cleinias makes the choice of wisdom with the idea that he is choosing it as the means to a popular and not a philosophical happiness, and thus perhaps gets more than he bargained for, but the probability that he understands εὖ πράττειν in the sense of prosperity rather than virtue has no connection with the argument; the discussion could have proceeded in the same way if he had taken εὖ πράττειν in the other sense.[6]

Another indication of the fact that Socrates is concerned with the things denoted by words rather than purely with the words themselves is his freedom in the use of synonyms. εὐτυχεῖν and εὐδαιμονεῖν are substituted for εὖ πράττειν, for instance, and σοφία is replaced on occasion by ἐπιστήμη, and even by φρόνησις. Apparently it is a matter of indifference what terms are used so long as the general meaning is clear; as far as the argument is concerned, wisdom can equally well be called knowledge or prosperity happiness. With the sophists the situation is quite otherwise. Because they are arguing to a strict contradiction, and because they are doing so by means of devices which require the use of single expressions in different senses, they cannot use synonyms in this free fashion for fear of upsetting the argument. Socrates can do so, first, because the things to which he refers are clearly the same even if he expresses them differently, and, second, because his purpose is different: he is not arguing to a contradiction but exhorting to a choice. His discourse, when compared with that of the sophists, shows

[6] Cf. Shorey's note on *Republic* I 353E in the Loeb edition, 1953: '[The equivocation on εὖ πράττειν] does not seriously affect the validity of the argument, for it is used only as a rhetorical confirmation of the implication that κακῶς ἄρχειν, etc. = misery and the reverse of happiness' (p. 105).

us something of interest about the nature of correct speech: the form of language is less important than its ability to indicate correctly the things the speaker desires to express.[7]

In addition to what may be inferred from this scene about language, there are at least two other significant points which emerge from the discussion. The first is that the acquisition of wisdom is not represented as merely one among various ways of obtaining happiness, but as the only way. A man will be happy if he has many good things, but none of the things which are ordinarily considered good turn out to be so unless we possess this one thing, wisdom, which is good in itself. Without the guidance of wisdom, other so-called goods are indifferent in value; if they are guided by ignorance, they may even be bad. Wisdom is at first included with all the other 'goods', but Socrates' pretense of having forgotten good fortune and then finding out that good fortune is after all the same as wisdom, has the effect of singling out wisdom for particular notice. When the two additional points are made, that a good is not a good unless it benefits us, and that nothing can benefit us unless we know how to use it, wisdom takes a definite place as the one intrinsically good thing among a group of others that are only derivatively good.

The other notable point which results from the discussion is that wisdom is declared by Cleinias to be teachable at 282D. This assertion is not presented as the outcome of any logical progression, but appears to burst upon Cleinias as a spontaneous result of the process of exhortation to wisdom to which he has just submitted. (Cf. the 'leaping spark' of *Ep.* VII, 341C.) It will be remembered that Socrates, in his question to Dionysodorus at 274D, implied that eristic was incapable of justifying its own function; its practitioners claimed to teach virtue but could not show that virtue was in fact teachable. At the same time it was suggested that some other art could do this, this other art presumably being dialectic. A demonstration of dialectic has now taken place, and we see that Cleinias, who was simply confused and helpless at the end of the first demon-

[7] Cf. *Protagoras* 358B, where Socrates says, 'for whether you say pleasant or delightful or enjoyable, my excellent Prodicus, or in whatever style or manner you may be pleased to name these things, pray reply to the sense of my question.' Cf. also *Charmides* 163D.

stration, emerges from this one with the firm conviction that wisdom, if not virtue, can be taught.

To say that wisdom can be taught is obviously not the same as to say that virtue can be, but the treatment of wisdom in what has just preceded suggests strongly that if wisdom is not the same as virtue, it is certainly very like it. Wisdom is the one sole good, for instance, as we see from 281E, and, when the future of Cleinias is spoken of, e.g., at 275A, 278D, 283A, it is said to be desirable not only that he should become wise, but that he should become virtuous and good. Wisdom and virtue thus appear to be linked in such a way that to aim at one is tantamount to aiming at the other. Whether the actual identity of the two in this dialogue be admitted or not,[8] the dialectical process has certainly yielded a result which is far more positive and striking than anything accomplished by eristic.

Scene III 282D–288D

Socrates now allows the sophists to continue their own demonstration, but not until he has, innocently enough it seems, laid down for them the direction he wishes it to take. That is, he presents them with the alternative of showing whether Cleinias ought to acquire every sort of knowledge or only one; in either case he assumes that they will want him to acquire *some* sort. Socrates continues his pretense of expecting to hear some wonderful arguments, and certainly gets them, in a way (θαυμαστόν τινα 283B). He again insists on his earnestness in the matter of the fate of Cleinias, but the sophists carry on in their previous vein; they do not exhort, but confute.

The arguments which the sophists now employ are again fallacious and are again constructed by means of the two devices used in their first performance, the fallacy of equivocation and the fallacy of *secundum quid.* Since the brothers are this time playing on certain ambiguities in the verb 'to be' (i.e., in later terminology, they are deliberately confusing its use as copula

[8] E. S. Thompson, in writing of *Euthydemus* 282C in relation to the *Meno* (The *Meno* of Plato, edited with Introduction, Notes and Excursuses, London, 1901, p. l), says 'The shifting of the subject of the question from ἀρετή to σοφία is not important'; he regards this whole passage as relevant to the question, is virtue teachable, as propounded in both the *Meno* and the *Protagoras.*

with its existential use), the results are much more serious from a philosophical point of view, as we shall see.

Dionysodorus begins 283B by asking whether Socrates and his friends really and truly desire Cleinias to become wise. They assure him earnestly that this is the case. But if they wish him to become wise and not to be what he is now, i.e., ignorant, they evidently wish him to become what he is not, and no longer to be what he is. In other words, they wish him not to *be*, i.e., to be dead and gone.

The mechanics of this first argument may be easily explained. Dionysodorus first obtains from Socrates the statement that he wishes Cleinias not to be ignorant. Then, by dropping the word 'ignorant', he asserts that what Socrates has really said is that he wishes Cleinias not to be; i.e., not to exist. He has used the fallacy of *secundum quid* in moving from not being some particular thing (ignorant) to not being absolutely, and has at the same time altered the sense of 'be', which when joined with 'ignorant' was used as copula, but which, when taken alone, naturally suggests existence.[9]

The argument appears trivial enough but is actually of considerable significance, since it is based on a metaphysics incompatible with Plato's, namely, that of Parmenides. The essence of Parmenides' philosophy was, of course, that what is is, and what is not is not; consequently, there is no becoming. From the Parmenidean point of view, if Cleinias is at the moment ignorant, as his friends admit, then he must remain ignorant, since this is what he *is*. If his friends say they wish him to become something else, they are really talking nonsense since nothing else exists for him to become. Therefore they can only mean that they wish for his non-existence. The clearest solution of this difficulty occurs in the *Sophist* (e.g., 257Bff.) where Plato shows that it is possible for one thing not to be another thing and yet still exist; it can be other instead of absolutely not. In this way he saves becoming and is consequently able to give some sort of reality to particulars. The Eleatic logic which

[9] On this passage see also E. H. Gifford, *The Euthydemus of Plato with Revised Text, Introduction, Notes, and Indices*, Oxford, 1905, who writes that at 283D, 'the pronoun ὅς is here equivocal, being used both in its proper sense as referring to a person and in an adjectival sense like οἷος' (p. 36). This may be true, but the fallacy does not depend on this equivocation alone.

is at the bottom of the majority of fallacies in this present scene is inextricably tied to a metaphysics which denies becoming; as a result it is necessary for Plato to expose these arguments, since in so doing he discredits the metaphysics as well. Until this work is done, he cannot provide a satisfactory basis for the theory of Forms. Here in the *Euthydemus*, the task is primarily that of exposure; the constructive work is accomplished elsewhere.[10]

Ctesippus is very annoyed at the affront to his favourite Cleinias, and upbraids Dionysodorus for speaking so falsely 283E. This gives Euthydemus the opening he needs for another fallacious argument, one which results in the denial of the

[10] E.g., *Republic* 476D ff. This passage in the *Republic*, which leads to the establishment of opinion as something midway between knowledge and ignorance 478D, is of exceptional interest, since the argument begins in a way which might well be the beginning of a sophistical argument. Socrates is considering what he and Glaucon might say to the man whom they have classified as having opinion, not knowledge. Socrates will ask questions and Glaucon is to answer on this man's behalf. The first question is (476E) 'Does he who knows know something or nothing?' 'He knows something,' Glaucon replies. 'Is it something that is or is not?' 'Something that is,' Glaucon answers, adding 'How could that which is not be known?' Socrates and Glaucon agree that what is is entirely knowable, and what in no way is is entirely unknowable. Up to this point the progress of the argument would be perfectly satisfactory to Euthydemus or Dionysodorus. These two would presumably proceed by bringing forward a case of something which is not some other thing, and then, by use of *secundum quid*, would argue that this thing is absolutely not, and thus is absolutely unknowable. Socrates is of course in agreement with them that the absolutely not cannot be known (he says so at 477A, as we have just seen) but he would disagree with them as to what things, if any, are to be placed in the category of the absolutely not. The sophists, who are working on the basis of a logic tied to Eleatic monism, would of course argue out of existence anything which is other than what is. (The thing which is may alter but is only one thing at a time.) Plato, who is concerned to recognize two orders of being, that of the Forms and that of the many participating in them (see above 476A) diverges completely from the Eleatic type of argument at 477A when he has Socrates say 'If a thing, then, is so conditioned as both to be and not to be, would it not lie between that which absolutely and unqualifiedly is and that which in no way is?' The Eleatic rule, of course, is that a thing must either be or not be, and there would, for a sophist of the Eleatic type, be no possibility of the existence of a thing which simultaneously is and is not. Plato has provided an object for his faculty of opinion in clear and conscious opposition to the Eleatic rule, and has, by implication, made a frontal attack on Parmenidean monism.

possibility of false speech. He persuades Ctesippus to say that a person who lies speaks the thing about which he is speaking, or, in other words, speaks that which is. But to speak that which is, or to speak things that are, is to speak the truth, so Dionysodorus cannot have been lying. Ctesippus objects, since whoever spoke as Dionysodorus did does not say things that are. But, continues Euthydemus, the things that are not surely are not, and can be nowhere; no one can possibly make them be in such a case. Now when orators speak in public do they do nothing? No, they do something, says Ctesippus. Then they also make, and speaking is a kind of making and doing. No one, therefore, speaks what is not, since he would then be making something, and Ctesippus has agreed that no one makes what is not. And if no one speaks what is not, no one speaks what is false; Dionysodorus therefore speaks what is true and what is. Ctesippus has to admit that Dionysodorus speaks what is, but he still protests that he does not speak it as it is.

Here the argument relies again on the fallacy of equivocation. The verb 'to be' (here τὰ ὄντα) is again taken in two senses, as meaning either 'to exist' or 'to be true'. Conversely, its negation is taken to mean either 'not to exist' or 'to be false'.[11] Therefore as soon as Ctesippus admits that the man who speaks speaks what is (by which he means only that the man speaks about a thing which exists) Euthydemus changes the sense of 'to be' and accuses him of having said that the man speaks the truth. When Ctesippus objects to this and says that Dionysodorus has spoken things that are not (by which he means things that are not the case, i.e., are false), Euthydemus again changes the sense of the verb and accuses him of having said that things that are not are (i.e., that what does not exist does exist). The sophist has, moreover, strengthened his position by insisting that speaking is a kind of making, by which he implies that speech produces some object. This enables him to suggest that while the man who speaks what is produces an object which exists, the man who speaks what is not is attempting to produce an object which does not exist. Since this is obviously impossible, he is in a strong position to deny the possibility of

[11] Gifford, *op. cit.*, p. 36, thinks, rather, that the equivocation is on λέγειν. But to bring off the complete argument it is necessary that τὰ ὄντα be equivocal as well.

speaking falsely. The real key to the matter, however, is his deliberate confusion of what is not true with what does not exist, a confusion which, when combined with the Eleatic denial of the real existence of what is not, leads irresistibly to the denial of falsehood.[12]

As a result of Ctesippus' continued insistence that Dionysodorus does not speak the things that are *as* they are (ὡς ἔχει), we have next (284Cff.) a brief argument turning on the ambiguity in this phrase and also upon the ambiguity of *κακῶς λέγειν*, which can mean either to say that a thing is bad or to speak ill of this thing. The gist of this argument is that if there are some persons, e.g., good men, who speak of things as they are (i.e., tell the truth), then, since bad things are bad (or, in the Greek idiom, *badly*), they will speak badly of bad things, i.e., speak ill of them. In answer to this, Ctesippus says that Dionysodorus had best beware of being included among bad men, lest the good speak ill of *him*. Euthydemus supposes that the good also speak greatly of the great and hotly of the hot. And frigidly of the frigid, returns Ctesippus. Ctesippus is turning quite abusive, says Dionysodorus. No, says Ctesippus, he is merely giving the sophist a friendly warning to refrain from saying that he, Ctesippus, wishes his best friends to be dead and gone.

At this point 285A, Socrates intervenes. If the two brothers really know how to destroy a wicked man and make him into a good one, then he and Ctesippus ought not to quarrel over a *word* (μὴ ὀνόματι διαφέρεσθαι). Socrates himself is willing to volunteer for such treatment; the sophists may do anything they like to him so long as they make him good. Ctesippus agrees,

[12] It is perhaps not immediately obvious how the fallacy of equivocation could become part of the stock-in-trade of a neo-Eleatic sophist, since this fallacy appears to be based on the fact that a word has many meanings rather than one. (It would be natural to connect this fallacy with pluralism rather than with monism.) However, although the fallacy is based on the use of the same term in at least two different senses, its success as an instrument of refutation depends on keeping up the pretense that the term has only one sense. E.g., when Cleinias has asserted that it is the wise who learn and then has to admit that it is not after all the wise who learn but the foolish, he would not think himself refuted unless he supposed that 'learn' had the same meaning throughout. The term is being used in two senses, as we have seen, and the solution of the fallacy consists in pointing this out.

and tries to make his peace with Dionysodorus. After all, he was only contradicting him, and contradiction is not abuse.

Socrates of course knows perfectly well that the sophists not only have no interest in making anyone virtuous, but could not do so even if they wished, since they have committed themselves to a metaphysics and a logic which is incompatible with change. I think that this denial of change or becoming on the part of the sophists explains why their style of argument can never succeed in showing that virtue or wisdom can be taught. Teaching is an activity which assumes that the pupil is capable of progressing from ignorance to knowledge, from vice to virtue, that he can, in fact, become other than what he now is. The eristic of Euthydemus and Dionysodorus makes no provision for teaching, learning, or any other kind of becoming, since it is so closely associated with the Eleatic logic that it remains confined to the static opposition between is and is not. If the sophists were really able, by some other means, to make a man good, their terminology would be of very minor importance, but the arguments they employ make it inconsistent for them to do any such thing.

The argument against contradiction which occurs next (285Dff.) is cut to the familiar Eleatic pattern. Ctesippus has said that he is not abusing Dionysodorus but contradicting him, whereupon Dionysodorus affects surprise that Ctesippus should think there is such a thing as contradiction. His argument against it is as follows: every thing has its own description (λόγος) which is as it is, not as it is not. No one speaks a thing as it is not, as they have already agreed. If both speak the description of the thing, they are speaking the same words and thus do not contradict; if neither speaks it, they do not deal with the thing at all, and thus do not contradict; if one speaks it and the other does not, only one of them gives the description and thus cannot be contradicted by the other, who gives no description. Again the combination of equivocation and *secundum quid* is at work: the man who speaks what is not, i.e., not what the other man speaks, is said not to speak at all. The qualification of 'is not' is dropped and the sense of the verb 'to be' is altered from copulative to existential.[13]

[13] With the argument against contradiction here in the *Euthydemus* cf. *Sophist* 236E: '*Str.* You see, Theaetetus, it is extremely difficult to understand how a man is to say or think that falsehood really exists and in saying this

Ctesippus is silenced by this latter argument, but not so Socrates. He has heard this sort of thing many times (the followers of Protagoras used it, as did others still earlier) and it always fills him with wonder since it seems to upset not only the views of others but itself as well. The substance of it seems to be that there is no false speaking; if one speaks at all, one must speak the truth. This is apparently meant to extend to thinking also, so that there is no such thing as false opinion, nor is there ignorance, nor are there any ignorant men. Is it Dionysodorus' real view that there is no such thing as an ignorant man, or does he just want to say something startling? Dionysodorus replies that Socrates is to refute him. But how can there be such a thing as refutation if nobody ever lies? There is no such thing, joins in Euthydemus, and Dionysodorus cannot have requested Socrates to refute him, since no one can request what is not. In response to this Socrates has what he says is a rather foolish question to ask: if there is no ignorance, there can never be any mistakes, and if there can be no mistakes, whatever in the world do the two sophists pretend to teach? Or did they not say that they could teach virtue to any man wishing to learn? How can Socrates be such an old simpleton as to bring up things that might have been said last year, says Dionysodorus, and yet be at a loss how to deal with what is being said at the moment. But, says Socrates, this phrase 'to be at a loss how to deal with it' is just the same as 'to be at a loss how to refute it'—what else can the phrase mean? Dionysodorus refuses to answer, but insists on questioning Socrates. He asks whether things that 'mean' (*νοεῖν*) have life when they mean, or do

not be involved in contradiction. *Th.* Why? *Str.* The statement involves the bold assumption that not-being exists. But the great Parmenides, my boy, from the time when we were children to the end of his life, always protested against this . . .'. Plato's attitude towards Parmenides is always a mixed one, as may be seen, e.g., from a familiar passage at *Theaetetus* 183E (one not always quoted at sufficient length): 'Parmenides seems to me to be, in Homer's words, "one to be venerated" and also "awful". For I met him when I was very young and he was very old, and he appeared to me to possess an absolutely noble depth of mind. So I am afraid we may not understand his words and may be still further from understanding what he meant by them; but my chief fear is that the question with which we started, about the nature of knowledge, may fail to be investigated, because of the *disorderly crowd of arguments* which will burst in upon us if we let them in . . .'. (My italics.)

lifeless things 'mean' also. Only the ones that have life, answers Socrates. Do you know any phrase that has life? No. Then why ask what my phrase means? Socrates admits he has made a mistake, if mistakes are possible, that is. If he was not mistaken, they cannot refute him; if he was mistaken, they were wrong in saying there is no such thing as a mistake. Thus their argument has not found out how to prevent itself from falling down in spite of the wonderful way it has of knocking down the arguments of others.

The passage just summarized (286B–288A) contains Socrates' first really direct attack on the sophists. The exact nature of this attack is, however, rather puzzling, since although Socrates appears to say (286C, 288A) that the argument employed is self-refuting, it is not altogether clear in just what way he regards the argument as involving its own refutation. The climax of his criticism (the 'clownish question' at 287A) emphasizes not the reflexive nature of the argument, but its inconsistency with the sophists' claim to be teachers. If Euthydemus and Dionysodorus have argued ignorance out of existence, they have obviously cut the ground from under their own professional feet, but to say that an argument which denies the possibility of ignorance is incompatible with the claim to teach is not at all the same as to say that this argument is self-refuting in its own nature. It is possible that Plato meant to imply that since the argument against refutation is itself based on a refutation (i.e., false speaking is argued against on the assumption that those who say that there is false speaking are themselves speaking falsely), therefore the argument is self-refuting. However, in default of any clear exposition of this point, it seems safer to assume that when Plato speaks of the argument as upsetting itself he has in mind simply the fact that the sophists, whose 'teaching' has turned out to consist of nothing but refutation, have made themselves look ridiculous by arguing that refutation does not in fact exist. From this point of view, the reference to Protagoras at 286C makes good sense, since that philosopher's assertion that man is the measure of all things exhibits a like incompatibility with the claim to teach. (Plato expounds this difficulty fully at *Theaetetus* 161Cff.)

The positive result of Socrates' criticism is, therefore, to discredit the teaching ability of the sophists once again. This amounts, of course, to a renewal of his criticism of eristic. Socrates has made it clear before (e.g., 278B) that one of his objections to the sophistic method is that the sophists deal not in things but in words. This contrast is certainly implied at 286Dff., when Socrates asks, 'Is it merely to save your statement, Dionysodorus, that you state it so—just to say something startling—or is it really and truly your view that there is no such thing as an ignorant man?' Dionysodorus' answer, 'But you . . . are to refute me', shows that he is not interested in discovering whether ignorant men do exist, but only in continuing the game of words. His attempt to continue it with the equivocation on 'mean' proves of course to be abortive since Socrates can now confront him with his inconsistency in accusing an opponent of making a mistake when he has just maintained that mistakes do not exist. The brothers are temporarily silenced, and Socrates, who is now top dog, can resume his ironic pretense that something fine will ultimately emerge from their discourse. Before permitting them another round, however, he resumes his own exhortation of Cleinias to wisdom and virtue.

Scene IV 288D–293A

At the end of their previous conversation 282D, Socrates and Cleinias agreed that wisdom ought to be pursued. This pursuit, called philosophy, is an acquiring of knowledge, and the knowledge will be of a profitable kind involving a union of making and knowing how to use the thing made. Hence it will be unlike the arts of lyre-making, flute-making, or speech-writing. Even Cleinias can see that those who write speeches do not always know how to use them and vice versa. This latter art cannot be the one to make us happy in spite of its wonderful way of charming a crowd. Socrates suggests 290B that the art of generalship may be what they are looking for. Cleinias objects to this, since the general, who is a hunter of men, does not know what to do with his quarry; he hands it over to the politicians just as the huntsman and the geometer hand theirs to the caterer and to the dialectician.

Crito is astonished that all this should have come from Cleinias, but Socrates is sure he heard this speech—it certainly was not spoken by Euthydemus or Dionysodorus. Socrates tells

Crito that he and Cleinias went on examining various arts until they came to the kingly art (ἡ βασιλικὴ τέχνη 291B). This one, which was the same as the art of the statesman, appeared to be the one to which all the other arts handed over their products to be rightly used. But, on considering the matter further, they found that while other ruling arts, such as medicine and agriculture, had some effect to show as a result of their rule, this art apparently had none. Even if they defined it as the art by means of which men are made good, they were at a loss to say in what respect men were made good by this art. It should of course transmit some sort of knowledge, but knowledge of what? Seemingly it can only give knowledge of itself. So Socrates and Cleinias found themselves no nearer than before to the discovery of what knowledge it is that makes men happy. In their distress they again called on the visitors to help them.

Socrates' investigation has ended in a familiar paradox: the highest good has been identified with knowledge, but this knowledge has turned out to have no other subject than itself. (This is the same difficulty which occurs e.g., in the *Charmides*, in the case of temperance.) Socrates professes to be defeated at the end of the discussion, but it is obvious that he speaks ironically. He has, in fact, reason to be pleased with what has occurred, if only because his exhortation has resulted in the emergence of Cleinias as an independent thinker. (Again, of course, Plato means to exalt dialectic at the expense of eristic.) Furthermore, something has been accomplished on the philosophical side as well.

In the earlier conversation between Socrates and Cleinias it became clear that wisdom, which was at first included in a class with other good things, was actually the only one among these goods which was good in itself; the others were good only when guided by wisdom. In other words, wisdom was lifted out of the class of things conventionally called good and given a special status of its own. The same is true of knowledge in the present passage. If I understand Plato rightly, he intends to say something like this: knowledge in the ordinary sense is knowledge *of* something, it has some easily recognizable product. This product is good, but its goodness is dependent on the goodness of knowledge as such. Knowledge as such has no identifiable subject-matter except itself simply because it comprises all 'knowledges'. Socrates' discovery, therefore, that the kingly or political art possesses the special character of having only itself as subject-matter is, I think, meant to imply that it is identical with knowledge in itself and that it really is the art for which he and Cleinias have been seeking. It will meet the requirements they have agreed upon, since it will benefit men and make them happy. It will do this, however, not by teaching them any specific skill but by imparting an intellectual and moral quality which will enable them to rule over the specific skills and which will make these good by virtue of their subservience to the knowledge which is itself the only thing good in itself.

Thus there is a sense in which Socrates is no more a teacher than the sophists are, since neither of them is teaching a particular subject. The sophists pretend to teach by indulging in verbal tricks which have no connection with matters as they really stand. (They have, in addition, made the teaching endeavor inconsistent for themselves by the positions they have adopted on becoming and false speaking.) Socrates likewise is not really teaching wisdom and virtue, he is exhorting to them. That is, he is not attempting to convey a body of knowledge, but rather trying to convert the answerer to a state in which that person will be able to find out the body of knowledge for himself. The word 'convert' is accurate, since what Socrates is working towards is a change of soul.

In the present scene, therefore, Socrates has not only confirmed Cleinias in his intention of becoming wise, but he has given him some indication of the sort of wisdom he must acquire. Cleinias will no doubt fail to grasp the total implications of what has been said, but he has at least been set on the right road, since he is now convinced that wisdom must be pursued and that, when he finds it, it will be something useful and good. As a result he should be well-equipped to reject the spurious substitutes which are being offered by Euthydemus and his kind.

Scene V 293A–304C

Euthydemus makes a pretense of coming to the rescue of Socrates and Cleinias but it is immediately obvious that he is up to his old tricks when he asks 293B 'Would you rather, Socrates, that I instructed you as to this knowledge . . . or propound that you have it?' As we are now accustomed to expect, he has no desire to talk about things but only about talking. The argument that follows again makes use of the fallacy of *secundum quid.* That is, from Socrates' admission that he knows one thing, Euthydemus moves to the conclusion that he is simply knowing (omniscient). Plato's mastery of this fallacy is complete, as may be seen from the way in which Socrates makes it difficult for Euthydemus to bring the refutation off. Socrates does this by adding qualifications to his answers, qualifications intended to prevent the sophist from moving from the particular to the general. For instance, when Euthydemus asks at 293B, 'Do you know anything?', Socrates replies, 'Yes . . . many things', and again, when Euthydemus says, 'Then you are knowing if you really know', Socrates answers, 'Certainly, in just that something' (*τούτου γε αὐτοῦ*). Socrates continues 293C to prevent Euthydemus from placing him in the category of either the knowing or the not knowing by insisting that while there are some things he knows, there are also many he does not know. When Euthydemus tries the opposite tack and wants to say that Socrates is absolutely not knowing, Socrates answers, 'Not in that thing' (*ἐκείνου*). Euthydemus then invokes the law of contradiction to show that Socrates cannot claim both to be and not be the same man (i.e. to be and not to be the man who knows) in regard to the same matter at the same time; since this is an impossible position, he must admit that he has been shown to know everything, including that knowledge he was seeking. Quite aside from the fact that the sophists have argued against contradiction earlier, this is wrong on two counts in the present context: (1) Socrates has not really allowed himself to be placed in either category, and (2) he has not said he did or did not know the same matters, but different ones. Socrates, however, does not trouble to make either of these objections. He contents himself with trying to place the sophists in the same predicament in which they have placed him, i.e., of being both knowing and not knowing. 'Tell me', says Socrates at 293E, 'you both know some existent things, of course, and others you do not?' Euthydemus and Dionysodorus will have none of this, since such an admission would destroy their position. Instead, they claim to know everything, as in fact do the rest of mankind, since 'they cannot know some things and not others, and so be at once knowing and unknowing' (294A). Socrates and Ctesippus press them to display their knowledge of some particular point, but this they decline to do, having no desire to go from the universal to the particular; they merely continue to repeat that they know everything.

As Socrates still pretends incredulity, Euthydemus embarks on a slightly different form of the same argument. Socrates, however, continues to qualify his answers as before, much to the sophist's annoyance. For instance, when Euthydemus asks whether Socrates knows 'by one and the same means always . . . or sometimes by one and sometimes by another' (296A), Socrates answers, 'Always, whenever I know, it is by this means'. Socrates is afraid, he says, that 'this word "always" may bring us to grief'. This is of course the crucial point, since as soon as Euthydemus can connect 'always' with 'know', he can set up his argument as before. This he proceeds to do, more or less over Socrates' dead body. He decapitates the phrase 'you know always by the same means' to 'you know always' (296C). Socrates lets this pass for a moment because he has an extremely damaging conclusion to draw from the premise with which he is confronted. He shows 296E that if he knows all things, then he must know not only true statements, such as 'good men are just', but also their (false) opposites, e.g., 'good men are unjust'. Dionysodorus, who is a little less wary than his brother, falls into the trap and asserts that Socrates cannot know this statement. He is reprimanded by Euthydemus, who sees that if Socrates can prove that there is one thing he does not know, they will lose their advantage. Socrates, meanwhile, is quick to add to their discomfiture by pointing out that it would be impossible for Dionysodorus to speak wrongly if he really knew everything.

The sophists wriggle out of this latter difficulty by picking up the word 'brother', which Socrates has dropped by the way, and are presently back at the old stand, happily engaged in confounding that which is other with that which is not. By means of this fallacy and by the use of equivocations on such words as 'his', 'yours', and 'mine', they establish in turn that Socrates has no father, that any man who is a father is father of all, and that Ctesippus' father is a dog. Then follows a brief argument 299B to the effect that if anything, e.g., medicine, or arms, or gold, is a good, then one ought to desire as much of it as possible in all times and in all places. (This is *a dicto simpliciter ad dictum secundum quid*; what is good absolutely is said to be good in some particular ridiculous way.) The mention of a talent of gold in one's skull 299E leads Ctesippus to pay the sophists back in their own coin by a play upon the ambiguity in the word 'theirs': the skulls belong to the Scythians, therefore when they drink out of them, they drink out of their own skulls. Euthydemus, taking advantage of the fact that Ctesippus has spoken of the Scythians as gazing at their skulls, begins in return to exploit the ambiguity in *δυνατὰ ὁρᾶν* 300A, which can (by reason of the Greek accusative and infinitive construction) be twisted to mean either 'possible to see' or 'able to see', and then in similar fashion to work with *σιγῶντα λέγειν*, which he takes at will as either a silent person speaking or as a speaking of something silent. Dionysodorus again shows himself to be somewhat less on his guard than Euthydemus by answering 'Neither and both' to Ctesippus' question 300D 'are all things silent or do they speak?' Ctesippus is now sufficiently familiar with the sophists' methods to seize upon the word 'both' as a slip, since of course the eristic tricks are largely dependent on the classification of everything into one of two exclusive classes. (Socrates, for instance, was said to be either knowing or not knowing, but not both.)

The brief passage 300E–301C which now follows is of unusual interest since it is the only place in the dialogue which appears to refer to the theory of Forms.

From the Eleatic point of view, the chief objection to the theory must be that it calls for two distinct sorts of existence, that of the Forms, and that of the particulars participating in them. Dionysodorus' question to Socrates at 300E, whether he finds beautiful things to be different from or the same as the beautiful, appears to be put with this objection in mind. If Socrates should answer that beautiful things are the same as the beautiful, Dionysodorus could argue that there is no distinction between them; if Socrates should answer that they are different, he could argue, by the usual fallacious equation of the other with the not, that beautiful things do not exist. In either case, monism would be preserved. Socrates, apparently perceiving the direction of the sophist's attack, attempts to answer the question in the terms in which it has been asked and yet to avoid representing the relationship between particulars and Forms as one of either total sameness or total difference: he says that beautiful things are different from the beautiful but that each has some beauty present with it (*ἕτερα . . . αὐτοῦ γε τοῦ καλοῦ· πάρεστι μέντοι ἑκάστῳ αὐτῶν κάλλος τι* 301A).

The notion of 'presence with' is an obstacle to the type of refutation the sophist presumably has in mind, since it provides some sort of mediation between beauty and beautiful things without, however, making them identical. He therefore attacks this notion in his next question, saying 'So if an ox is present with you, you are an ox, and since I am present with you, you are Dionysodorus'. This is interesting on two counts, first that Dionysodorus has taken presence in the purely physical sense, and second that he has destroyed the contrast between Forms and particulars by dealing in terms of particulars only. The two points are interrelated and are both connected with monism. Dionysodorus, in order to confine himself to one sort of existence, has spoken not of oxness as present with Socrates (as he ought to do if he is to meet the theory of Forms on its own ground) but of an ox as so present. Since both Socrates and the ox are physical particulars, the only possible mode of presence is physical presence. Dionysodorus has made Socrates' answer look ridiculous, but he has been able to do this only by distorting the nature and sense of Socrates' terminology.

Dionysodorus now follows up the joke about the ox by phrasing what is basically the same objection in another and more subtle way. He says 301A 'In what way can one thing, [i.e. the different] by having a different thing present with it, be itself different?' (*Ἀλλὰ τίνα τρόπον . . . ἑτέρου ἑτέρῳ παραγενομένου τὸ ἕτερον ἕτερον ἂν εἴη;*) In this new question he is still attacking the idea of 'presence with', but is at the same time taking advantage of Socrates' statement at 301A that beautiful things are different from the beautiful itself. His procedure is to disregard what it is that both the beautiful and beautiful things are different *from* (each other) and to refer to both as simply 'different', or, if it suits his purposes better, as '*the* different'. This is the familiar fallacy of *secundum quid* at work, and the result of the manoeuvre is again, as in the case of

Socrates and the ox, to destroy the distinction between Forms and particulars. We can see how this has happened by putting down the sophist's question with its correct expansions. It should read, 'In what way, because of beauty being present with beautiful things, can the beautiful things be beautiful?' In its truncated state, however, it has become, in effect, 'How can the different be different?' The distinction between beauty and beautiful things has been swallowed up, and with it the distinction between two sorts of existence. Furthermore, the question is now no longer one concerning the mode of participation of particulars in their homonymous Form, but one which inquires about the possibility of a Form being predicated of itself. The Form about which this question is asked is one which the sophist has constructed by the transformation of the adjective 'different' into an adjectival noun, 'the different'. This 'Form' he apparently assumes to have the same status as the beautiful, so that he raises no objection when Socrates argues that if the beautiful is beautiful and the ugly ugly, then, by analogy, the different will be different and the same the same. Socrates has given a strong hint that he is refuting the sophist by fallacious means since he says at 301B, 'already I was attempting to imitate the cleverness of these men, I was so eager to get it'. We have reason, therefore to question whether he regards the statement that the beautiful is beautiful as a legitimate one. It seems to me likely that he does not, since this statement is so clearly modelled on the sophist's truncated question about the different.[14] However, the legitimacy of this statement, though an interesting topic in its own right,[15] is not

[14] With this passage cf. *Sophist* 254Dff. and A. L. Peck, 'Plato and the *ΜΕΓΙΣΤΑ ΓΕΝΗ* of the *Sophist*', *Classical Quarterly*, 1952, pp. 46–9.

[15] For a discussion of the self-predication of Forms in Plato, see G. Vlastos, 'The Third Man Argument in the *Parmenides*', *Philosophical Review*, LXIII, 1954, pp. 319–49. (See also articles subsequent to this paper, a bibliography of which may be assembled by consulting R. S. Bluck, 'Forms as Standards', *Phronesis*, 2, 2, 1957, pp. 115–27, and N. B. Booth, 'Assumptions involved in the "third man" argument', *Phronesis*, 3, 2, 1958, pp. 146–9. To this list should also be added the excellent paper of R. E. Allen, 'Participation and Predication in Plato's Middle Dialogues', *Philosophical Review*, LXIX, 1960, pp. 147–64.) A number of scholars, of whom T. A. deLaguna is perhaps the first ('Notes on the Theory of Ideas', *Philosophical Review*, XLIII, 1934, p. 450), have also noticed that an instance of the self-predication of Forms

the main point at issue. The point is that Socrates himself has used an analogy between this statement and the illegitimately constructed statement about 'the different' to bring about what is not a refutation of the sophist's original objection to Socrates' account of the relation between Forms and particulars, but a refutation of the distorted shape in which this account now appears, and he has done this by means of exactly the same distortion.

---

occurs at *Protagoras* 330C. (See Vlastos, *op. cit.*, p. 337, who calls this 'the star instance of Self-Predication in Plato'.) I feel myself that Socrates, when he asks Protagoras at 330C, 'Is justice a thing or not' (ἡ δικαιοσύνη πρᾶγμα τί ἐστιν ἢ οὐδὲν πρᾶγμα), is deliberately misleading the sophist, and that he asks the question in order to persuade Protagoras to place justice in the same class as particular just acts. When Protagoras does this, it then appears reasonable, as it would not perhaps otherwise have done, to inquire whether this thing, justice, is itself just or unjust. This whole passage needs to be studied in the general context of the reduction of the various virtues to knowledge, and of the ultimate refutation of Protagoras, who has attempted to maintain both that the virtues are distinct 330A and that virtue can be taught (317B, 319B). What Socrates does in the dialogue is to show that Protagoras' claim to teach virtue implies not, what he has also asserted, that the virtues are distinct, i.e., many, but that they are one, i.e., that all can be reduced to knowledge. The argument which employs the statement 'justice is just' is the first step in the reduction of the virtues. By this means, and by means of some additional fallacies such as the passage from 'not just' to 'unjust' (331A), two of the virtues, justice and holiness, are equated. Socrates then proceeds to equate two others, temperance and wisdom. This he does by a clearly fallacious argument: the term 'folly' is used in one sense as opposite to temperance and in another sense as opposite to wisdom—the term is employed equivocally, in other words. Socrates is about to bring together the two complexes justice-holiness and wisdom-temperance by a third argument (which will apparently make use of an equivocation on 'temperate') when Protagoras, presumably detecting the direction of the argument, interrupts with his speech on the relative profitableness of things at 334Aff. The main lines of the argument are not picked up again until after the Simonides interval, i.e., not until 349B. The remainder of the dialogue is then devoted to the more difficult task of the reduction of courage to knowledge. This argument also appears to be fallacious, and consciously so, but it is too long to be analysed here. (See G. Vlastos' Introduction to Plato's *Protagoras* in the *Library of Liberal Arts* series, New York, 1956, p. xxxi. Professor Vlastos would not agree with me that the fallacies are deliberate, however.) The fact then that the argument employing the statement 'justice is just' appears as one step in a series of, as I think, consciously fallacious arguments, leads me to suspect strongly that this argument is consciously fallacious also. Furthermore the present passage in the *Euthydemus* strikes me as confirmatory evidence.

To sum up: Dionysodorus has attacked the theory of Forms by means of his accustomed weapons of equivocation and *secundum quid*, and has, as a result, completely misinterpreted the theory because the contrast between Forms and particulars has disappeared. Socrates has defended himself against the attack, not by pointing out the fallacies involved (as he does, for instance, in the case of the equivocation on 'learn' in the first scene), but by accepting the misinterpretation of the theory and then proceeding to refute the sophist in the sophist's own terms. He has made him pay the penalty for constructing an illegitimate Form (the different) by forcing him to accept an analogy between it and legitimate Forms like the beautiful and the ugly. In a sense the theory, in its correct version, has not been defended at all, but Socrates has at least demonstrated that this sort of fallacious attack can be disposed of on its own grounds. He seems to me also to have demonstrated the complete incompatibility of the theory with Eleatic monism in a very striking way, since the fallacies which the sophist employs are Eleatic in origin and their effect upon the theory of Forms is to destroy its essential dualism.[16]

Of additional interest in connection with this passage is the fact that it is here that the sophists have come nearest to being philosophers. The relationship between Forms and particulars is undoubtedly the most difficult aspect of the theory of Forms and, in selecting this for attack, they show an instinct for a genuine philosophical problem. Then the ox joke, although based on a misinterpretation of the theory, does at least draw attention to the fact that one meaning of 'present with' is physical presence, and that Socrates has not taken this into account. Furthermore, by implying that in some way the presence of the ox with Socrates *causes* Socrates to become an ox, they have indirectly raised the question whether the beauty of beautiful things is caused by the presence of beauty with them. This is again a genuine philosophical problem, and one not touched upon by Socrates. Even the fallaciously constructed

[16] If Dr. A. L. Peck is right in his interpretation of the first part of the *Parmenides* (I think he is) this same incompatibility is demonstrated even more strikingly in that dialogue. See 'Plato's *Parmenides*: Some Suggestions for its Interpretation', *Classical Quarterly*, N.S., vol. III, Nos. 3 and 4, July-October, 1953, pp. 126–50.

question, how can the different be different, involves two interesting philosophical points: can a Form be made from a relational adjective, and can a Form be predicated of itself? I think that Plato, in this passage, means to point out the very thin partition between philosophy and sophistry: the sophists have raised some important philosophical issues, but their love of refutation and word-play is so strong that they are incapable of investigating any of them seriously.

The argument about cutting up the cook which follows 301D is not of great interest, since it is merely another instance of a fallacy of the same sort as those involving the phrases δυνατὰ ὁρᾶν and σιγῶντα λέγειν (i.e. it is again made possible by the Greek accusative and infinitive construction). This is followed 301E by another argument based on the equivocal use of 'mine' and 'yours' which places Socrates in the position of claiming to sacrifice his ancestral gods 303A. 'Bravo, Hercules . . . a fine argument,' exclaims Ctesippus. 'Now do you mean that Hercules is a bravo, or that bravo is Hercules?' asks Dionysodorus. At this there is wild applause from all sides. Even Socrates himself joins in praise of the clever pair who have acquired such a great accomplishment in such a short time. Obviously they care nothing for the multitude, but only for those of their own sort, since most people would consider it a greater disgrace to refute with such arguments than to be refuted by them. Also they show very good manners in being willing to stitch up their own mouths as well as those of others.[17] Best of all is the way in which anyone else may learn this skill so quickly, as Ctesippus has been observed to do. However,

[17] The passage 303D reads 'when you say there is nothing either beautiful, or good, or white, and so on, and no difference of things at all, in truth you simply stitch up men's mouths' and so on. The sophists have not explicitly maintained these propositions in the dialogue, so that it is difficult to see to what Socrates refers. My guess is that this is an oblique reference to the fact that the sophists, in their attack on the theory of Forms 301 ff., have, because of their monism, denied the existence of beautiful things. Thus they must, to be consistent, deny the existence of all other sorts of particulars, good things, white things, etc. Their monism of course leads them also to the denial of the existence of the difference of things, since if any one thing is, any other thing, differing from it, does not exist. The denial of false speaking earlier in the dialogue can also be regarded as a denial of the difference of things, since it destroys the distinction between truth and falsity. (See below on the *Cratylus* pp. 48 ff.)

they will be wise not to use it in too many public discussions in case other people should learn the whole thing straight away and give them no credit for it. They should talk only to each other, or, if anyone else is present, be sure to charge him a good fee. (With this speech of Socrates, the related portion of the dialogue ends.)

Epilogue 304C–end

Socrates again suggests to Crito that they should both go to school to the sophists. Crito is willing, but fears he is one of those who would prefer to be refuted by such arguments rather than to use them himself. He wants to tell Socrates of a conversation he has just had with a man who heard the discussion and who, as a result, was very critical not only of philosophy but also of Socrates, for being willing to take part in such a performance. Socrates suspects that this was one of those clever persons, half-way between philosophy and politics, who write speeches for the law-courts. This man feels that Euthydemus and the like stand in his way, and thus seeks to decry philosophy in order to enhance his own reputation. He ought to realize that a person midway between two good things is inferior to both of them, and that he really occupies the third place, not the first. Still we should be grateful for anyone who says anything bordering on good sense. Crito is much in doubt what to do about his sons, when he sees the sort of persons who profess to educate people. But, says Socrates, he should not be deterred from philosophy simply because there are some bad teachers in the business. He should test it himself, and if he finds it to be what Socrates maintains it is, then he should pursue it, he and his children too.

To decide whether Isocrates or some other is referred to in this last section seems to me not of great importance in so far as the general meaning of the passage is concerned. Since Socrates, in being 'so strangely willing to lend himself to persons who care not a straw what they say, but merely fasten on any phrase that turns up' (305A), does not seem, to the casual observer, to differ greatly from the sophists, it is time for a defence of philosophy, both as an occupation and as a method of instruction. The criticism made by the semi-philosopher semi-politician serves to provide the occasion for this defence.

The defence itself, however, is curiously indirect. It begins by exposing the inconsistency in the speech-writer's position. According to the principle laid down by Socrates at 306Aff. this man can only be right in regarding himself as superior to both the philosopher and the politician if philosophy and politics are both bad. If both are good, he is inferior to both; if one is good and the other bad, he is worse than one, but better than the other. It is obviously to his advantage to malign both professions, but, according to Socrates at 306C, he is apparently unwilling to do this. Thus, we are intended to conclude, his attack is inconsistent.

Since Crito, at 305A, has just quoted the speech-writer as saying that both philosophy and the people who follow it are worthless and ridiculous, it seems rather strange that Socrates should represent the man as declining to admit that either philosophy or politics is bad. The only possible explanation of this seems to be that Socrates is trying to bring out the fact that the speech-writer, must, in his own mind, be making some distinction between a bad philosophy practised by the sophists and some other pursuit, also called philosophy, which he regards as good. This explanation is borne out by what occurs at the very end of the dialogue 307Aff. since here Socrates advises Crito not to eschew philosophy because some, or even most, of its practitioners are worthless. (We can think of two he has in mind.) What Socrates has been trying to achieve in the rather roundabout preliminaries to this final point is a separation between the genuine philosophical activity and its false imitation. This he does by bringing forward the case of a man who attacks philosophy and yet is unwilling to admit that philosophy is bad. The inference, which is left for us and for Crito to draw, is that what was being attacked under the name of philosophy was really eristic. Philosophy in its proper sense remains a good, and as such, is unaffected by the practises of those who pretend to engage in it but are in fact doing something else. Crito must of course test the matter for himself and make the decision about his own education and that of his sons accordingly, but we are certainly left with the impression that he can hardly fail to decide in favor of philosophy as Socrates conceives it. The alternating demonstrations of dialectic and eristic have had their effect upon him as well as upon Cleinias, and the discussion has accomplished its purpose even at second-hand. It is Plato's obvious intention that it should accomplish its purpose with the reader as well.

The foregoing summary of the *Euthydemus* has shown that the results of the dialogue are chiefly two: the recommendation of philosophy conceived as dialectic (since it is by this means that an individual may be exhorted to the choice of wisdom and virtue), and the rejection of philosophy conceived as eristic. Plato's main reason for rejecting eristic is that it disregards correct speech and thus gives no reliable information about the true state of things.[18] Euthydemus and Dionysodorus have treated words as counters in a game; hence their verbal tricks can convey nothing as to how matters really stand, and, as Plato demonstrates rather than asserts, can lead to no alteration in the individual soul. The rejection of eristic is made especially crucial for Plato by two circumstances: first, that the sophists propose to teach virtue by this means, and second, that their arguments are for the most part based on a metaphysics (that of Parmenides) which is incompatible with his own. In exposing the fallacious nature of these arguments, therefore, Plato is not only defending his own method of teaching but also clearing the way for what is, after all, the indispensable condition of this method, the theory of Forms.

I now propose to consider, in the next two chapters, some fallacious arguments similar to arguments in the *Euthydemus*. As already stated in the introduction, I believe this similarity to warrant the assumption that their employment by Plato in these other dialogues is as deliberate as it clearly is in the *Euthydemus*. The philosophical issues involved will again be of considerable significance, as we shall see.

[18] This does not mean, contrary to what is said in the *Cratylus*, that Plato wants to say here that correct speech leads to knowledge of the nature of reality. What he does want to say, I think, is that we should employ the best terms we have, without perverting their accepted meanings.

CHAPTER TWO

Theaetetus 163Aff.

VERY early in the dialogue, at 145E, Socrates asks the young Theaetetus for a definition of knowledge. Like various other persons questioned by Socrates, Theaetetus at first fails to grasp the nature of general definition, but he soon comes to understand what Socrates wants, and, at 151E, he puts forward the view that knowledge is perception. Socrates' reaction to this is to point out that such a definition is tied both to the relativism of Protagoras and to the all-flowing philosophy of the Heracliteans. He proceeds to develop a number of consequences of the flowing philosophy, but does so in such a way that Theaetetus is not clear as to whether this philosophy (and thus his own definition) is being attacked or not 157C. Even after the passage 157Eff. in which the flowing philosophy turns out to provide no criterion by which waking may be distinguished from dreams or sanity from insanity, Theaetetus is apparently still in doubt, since he raises no objection when Socrates says 160D:

Therefore you were quite right in saying that knowledge is nothing else than perception, and there is complete identity between the doctrine of Homer and Heraclitus and all their followers—that all things are in motion, like streams—the doctrine of the great philosopher Protagoras that man is the measure of all things—and the doctrine of Theaetetus that, since these things are true, perception is knowledge.

But after the speech 161Cff. in which Socrates suggests that Protagoras' measure might as well be a pig or a dog-faced baboon, Theaetetus' faith in his definition is severely shaken; he says 162C, 'when we were discussing the meaning of the doctrine that whatever appears to each one really is to him, I thought it was good; but now it has suddenly changed to the opposite.' It is at this point, when Theaetetus has begun to wonder whether his definition will hold after all, that the passage with which I am concerned in this chapter begins.

'Well then,' Socrates says at 163A,

> let us look at it in this way, raising the question whether knowledge is, after all, the same as perception, or different. For that is the object of all our discussion, and it was to answer that question that we stirred up all these strange doctrines, was it not?

He then proceeds to give a set of three arguments, all of which show that the identity of knowledge with perception can lead to absurdity. The first two arguments play upon the ambiguity in the verb 'to know' (ἐπίσταμαι) in a manner similar to that in which the arguments in the first scene of the *Euthydemus* played upon the ambiguity in the verb 'to learn' (μανθάνειν).[19] The third argument employs the 'knowing—not knowing' dichotomy which occurred in the *Euthydemus* in Scene V as the basis of the sophists' claim to be omniscient, and which is also implied in the arguments of the first scene, as Socrates observes at 278A.

The first argument (which has two parts) takes 'know' in the sense of 'perceive' and also in the sense of 'learn'. It begins by using the example of a person hearing a foreign language he has not yet learned. If knowledge is really the same as perception, then when this person hears the language, he does in fact know it, or else, if he does not know it, he does not in fact hear it. Both conclusions fail to fit the facts as originally described. The same argument is then repeated for seeing: if we see letters we have not yet learned, we either know them after all because we see them, or we do not really see them because we do not know them. In the case of both the foreign language and the letters, an example has been given of a kind of knowing which is obviously not co-extensive with perceiving. The point is, of

[19] 'Know' and 'learn' were terms which were notoriously productive of sophistical arguments. Cf. *Theaetetus* 199A, ὅπῃ τις χαίρει ἕλκων τὸ ἐπίστασθαι καὶ μανθάνειν.

course, that although the word 'know' is not entirely unmeaningful in connection with hearing or sight, there is at least one other sense of 'know' with which the idea of perception is incompatible, and if these two meanings are combined, it is possible to construct contradictions. The two sets which have been given may be readily schematized as follows:

A (1) I hear German = (3) I perceive German = (5) I know German.
(2) I have not learned German = (4) I do not know German = (6) I do not perceive German.

B (1) I see Greek letters = (3) I perceive Greek letters = (5) I know Greek letters.
(2) I have not learned Greek letters = (4) I do not know Greek letters = (6) I do not perceive Greek letters.

(The contradictions are double, i.e., between (3) and (6), and (4) and (5), in each case.)

Theaetetus' answer at 163B, that 'we know just so much of [the language and the letters] as we hear or see . . . [but we do not know] what the grammarians and interpreters teach about them', shows that he understands perfectly that it is the double use of 'know' which has produced the contradictions.

The second argument shows that there is still another sense of 'know' which is not accounted for if knowledge is said to mean simply perception, i.e., the sense of remembering what has been learned. The question is asked at 163E whether a person remembers, i.e., knows, when his eyes are shut, i.e., when he is not perceiving. Theaetetus answers that it would be absurd to deny such a thing, so that we again have a case in which the identity of knowledge and perception has resulted in a paradox:

C (1) I remember x = (3) I know x = (5) I perceive x.

D (2) I do not see x = (4) I do not perceive x = (6) I do not know x.

(The contradictions are between (3) and (6), and (4) and (5), as before.)

In the first argument, it turned out to be possible to perceive without knowing; now it turns out to be possible to know without perceiving. Knowledge and perception cannot therefore be said to be the same, but we must say that they are different 164B.

This conclusion is not allowed to stand for more than a moment, however, since Socrates immediately objects 164C, D, that he and Theaetetus have been 'acting like professional debaters' (ἀντιλογικῶς) and have 'based [their] agreements on the mere similarity of words' (τὰς τῶν ὀνομάτων ὁμολογίας); they 'do not see that [they] who claim to be, not contestants for a prize (ἀγωνισταί), but lovers of wisdom (φιλόσοφοι) are doing just what those ingenious persons do' (τοῖς δεινοῖς ἀνδράσιν). Furthermore, if Protagoras had been alive, he would not have allowed the definition to be treated in such a fashion. Socrates, therefore, undertakes to come to Protagoras' assistance, 'for a man might find himself involved in still worse inconsistencies than those in which we found ourselves just now, if he did not pay attention to the terms which we generally use in assent and denial' (165A). We then have a third argument against the definition, which is also eristic,[20] and which Socrates describes as 'the most frightfully difficult question of all' (165B).

Socrates begins by asking Theaetetus whether the same person can both know and not know that which he knows. This, Theaetetus says, is obviously impossible. But suppose someone were to cover one of your eyes with his hand, and then to ask if you saw his cloak with that eye? Theaetetus answers that he would say he did not see the cloak with that eye but did see it with the other. Then, says Socrates, your opponent would conclude that you both see and do not see at the same time, and since, by the definition, seeing is the same as knowing, that you also both know and do not know at the same time. Theaetetus objects that his seeing and not seeing was only 'after a fashion' (οὕτω γέ πως 165C), but this qualification would not be allowed, according to Socrates. Furthermore he goes on to show (165Dff.) that the identity of knowledge and perception could give rise to a host of similar difficulties, any one of which, if

[20] It might be well to point out that I have usually employed the terms 'eristic', 'sophistical' and so forth in relation to arguments similar in form to those used by Euthydemus and Dionysodorus even when such arguments are not used contentiously, since, for my purposes, it was necessary to find some convenient way of indicating the similarity between arguments in the *Euthydemus* and arguments elsewhere. (Incidentally I argue below, pp. 43–45, that the fact that Socrates employed sophistical arguments in the *Theaetetus* does not in the least imply that he did so in a sophistical spirit.)

employed by 'a nimble fighter, fighting for pay in the war of words', would be sufficient to reduce Theaetetus to a state of contradiction.

Since Socrates has come to the aid of Protagoras and his definition by demonstrating that still another paradox may be generated by means of the very definition Protagoras would like to defend, we may reasonably regard this as a rather odd type of assistance. His offer is ironic, as we may have guessed from his remark to Theodorus at 165A: 'Now see how I shall help him' (*σκέψαι οὖν τήν γ'ἐμὴν βοήθειαν*). Nevertheless, the very perverseness of his procedure has the effect of bringing into prominence the objections which he attributes to Protagoras following the argument at 166Aff. The substance of all these objections is that perception has been conceived in a way which is too static.

Do you suppose you could get anybody to admit that the memory a man has of a past feeling he no longer feels is anything like the feeling at the time when he was feeling it? Far from it. Or that he would refuse to admit that it is possible for one and the same person to know and not to know one and the same thing? Or if he were afraid to admit this, would he ever admit that a person who has become unlike is the same as before he became unlike? In fact, if we are to be on our guard against verbal entanglements, would he admit that a person is one at all, and not many, who become infinite in number, if the process of becoming different continues?

What Plato seems to be having Protagoras point out is that it is neglect of becoming which gives rise to fallacious arguments of the type just used. Plato means, I think, to imply that the Heracliteanism with which he has associated Protagoras gives at least a hint of the correct way of avoiding such fallacies, although a completely flowing philosophy would err too much in the opposite direction. (This last he has already suggested in the passage 158Eff. in which Socrates ill turns out to be wholly other than Socrates well.) The fallacies which have been set up have all been constructed on an Eleatic basis (e.g., it is assumed that 'know' has only one meaning, whereas it in fact has several) and this fact is thrown into relief by their occurrence in a context of Heracliteanism. The first set of arguments displays the fact that the definition of knowledge as perception leaves out of account the phenomenon of learning; in the second argument we see that it also fails to account for memory. The third argument, in which knowing and not knowing are regarded as exhaustive alternatives, neglects the possibility of any mediation between the two extremes, or, in more Platonic language, it fails to distinguish the other from the not. Protagoras' objections have of course served to draw attention to the fact that these various forms of becoming have been neglected, but they do not, nevertheless, suggest the correct solutions; for these we must go to the comments given to Theaetetus.

In both parts of the first argument, Theaetetus appeared to be contravening the sophistic rule that it is impossible for the same person both to know and not know the same thing at the same time. He escaped from this difficulty by the distinctions made at 163B, the implications of which were that it is perfectly possible to know in the sense of perceiving and at the same time not know in the sense of not having learned. In the second argument, Theaetetus again has to go against the sophistic rule about knowing and not knowing, but the explanation is again that 'know' has been used in two senses: it is possible to know in the sense of remembering what one has learned and at the same time not to know in the sense of not at the moment perceiving this thing. Socrates' remarks about people who behave ἀντιλογικῶς are directed towards the way in which the arguments have been set up, and towards the fact that Theaetetus has met them by asserting, in effect, that it is after all possible to know and not know at the same time. Theaetetus has of course taken this position with the understanding that 'know' is an equivocal term, but Plato wishes to demonstrate that a sophist would ignore this: i.e., he would use the equivocation, but would conceal the fact. Socrates therefore asks Theaetetus outright whether it is possible for a person both to know and not know that which he knows, and to the question put in this bald form, Theaetetus of course says no. Plato then employs the technique already illustrated from the *Euthydemus* of dropping the qualifying part of a phrase (the fallacy of *secundum quid*). That is, although Theaetetus wishes to say (correctly) 'I do not see with the eye which is covered, but I do see with the eye which is uncovered', Socrates points out that his sophistical opponent would interpret this as an assertion of seeing and not seeing. Since seeing has been previously identified with knowing, Theaetetus is placed in the position of having contradicted his original statement that simultaneous knowing and not knowing is impossible.

The third argument thus serves a purpose slightly different from that of the first two. In (1) and (2), we have seen that Plato is not only pointing out that the identification of knowledge with perception may give rise to sophistry because 'know' is an equivocal term, but also that the definition does not account for two important psychological facts: learning and memory. (Since both are matters of becoming and degree, they are bound to clash with the sophists' rule of either-or.) The third argument has the more limited purpose of showing that 'knowing or not knowing' is itself a source of sophistry, but only so when put in its extreme form. Theaetetus has been using it correctly in (1) and (2); when it is formulated in the sophists' way, in (3), he quite rightly denies its validity.

In connection with this interpretation of the three arguments, it is important to take note of some additional passages later in the dialogue.

At 186E, after Socrates has made the distinction between sensations and reasoning, the definition of knowledge as perception is finally given up altogether. Theaetetus then offers a new definition, that knowledge is true opinion (*ἡ ἀληθὴς δόξα* 187B). He has not, he says, defined it as simply opinion, since there exists false opinion as well. Socrates immediately undertakes to examine, not the new definition, but Theaetetus' statement about the existence of false opinion. The way in which he begins the examination is very instructive:

This then, at any rate, is possible for us, is it not, regarding all things collectively and each thing separately, either to know or not to know them? For learning and forgetting, as intermediate stages, I leave out of account for the present, for just now they have no bearing upon our argument.

Th. Certainly, Socrates, nothing is left in any particular case except knowing or not knowing it. (188Aff.)

As soon as the 'knowing—not knowing' dichotomy is made operative, in other words, all forms of becoming must be dropped. This was so in the case of the sophistical arguments which Socrates brought against the 'knowledge is perception' definition, and is so again here when Socrates prepares to bring a similar type of attack against the existence of false opinion.

The attack against false opinion consists of two parts. The first is based on the assumption just quoted, that to know an object implies knowing it completely and infallibly and not to know an object implies knowing nothing about it whatsoever. (There is no room for any intervening degrees of knowledge, in other words.) On this assumption there can be no explanation of false opinion since, as Plato shows by the method of exhaustion, such opinion cannot arise through a man thinking either (1) that things he knows are other things he knows, or (2) that things he does not know are other things he does not know, or (3) that things he does know are things he does not know, or (4) that things he does not know are things he knows (188Bff.).

The second part of the attack is introduced by Socrates' suggestion that he and Theaetetus should substitute 'being and not being' for 'knowing and not knowing'. This of course paves the way for an argument of the type which the sophists employed against false speaking and contradiction in the *Euthydemus* 283Eff. The similarity shows clearly at the end of the argument 189A:

Soc. So, then, does not he who holds an opinion hold an opinion of some one thing?

Th. He must do so.

Soc. And does not he who holds an opinion of some one thing hold an opinion of something that is?

Th. I agree.

Soc. Then he who holds an opinion of what is not holds an opinion of nothing.

Th. Evidently.

Soc. Well then, he who holds an opinion of nothing, holds no opinion at all.

Th. That is plain, apparently.

Soc. Then it is impossible to hold an opinion of that which is not, either in relation to things that are, or independently of them.

Th. Evidently.

However, although Socrates has gone through two eristic arguments, he does not draw the typical eristic conclusion.

That is, instead of saying that false opinion does not exist, he concludes that it 'is something different from holding an opinion of that which is not' (189B). Plato is still determined to show that there is such a thing as false opinion (the resulting absurdities would otherwise be too numerous 190E and cf. *Sophist* 241D, E) and he has rehearsed these arguments to show that they are not really effective against it. This we have had a reason to expect in advance, since we have been warned ahead of time that becoming will be left out. After the rejection of another possible argument against false opinion (which takes the form of a demonstration that it cannot be ἀλλοδοξία or interchanged opinion), Socrates changes the direction of the discussion: he stops his negative defense of false opinion and shifts to a positive defense, that is, to the images of the wax tablet and the aviary. It is at this point that learning 191C and presently memory 191D are reintroduced. It therefore seems clear that in this part of the dialogue as in the passage 163ff. there is a direct connection between the presence of sophistry (of a particular type) and the absence of becoming.[21] (What I have loosely called becoming includes two things in the present context: (1) the psychological phenomena of learning, forgetting, and memory, and (2) degrees of knowing and not knowing.) This is presumably what Socrates has in mind when he says 157B that 'anyone who by his mode of speech makes things stand still is easily refuted'. But, as has been said earlier, a situation of total becoming would be, if not the producer of similar sophistical arguments, at least the producer of similar results, since, as Socrates has pointed out at 170C, the doctrine that man is the measure also abolishes false opinion. The reasons are different but the outcome is the same.[22]

[21] Euthydemus and Dionysodorus would never have defined learning as 'growing wiser about that which one learns', as Socrates does at 145D, for instance. It is interesting to note also that at *Sophist* 259C a criticism of the kind of quibble which results in the 'complete separation of each thing from all' precedes the introduction of the εἰδῶν συμπλοκή at 259E. In order to get rid of Eleatic sophistry, in other words, some device must be employed (here κοινωνία rather than the introduction of becoming) in order to avoid the extreme type of separation which is 'the utterly final obliteration of all discourse' (259E).

[22] By the same token we find that in the *Cratylus* the two oppposed theories of language (one Eleatic and one Heraclitean) really come to the same thing

The arguments in the *Theaetetus* 163ff. are, it seems clear to me, very similar to arguments in the *Euthydemus*, and thus I have conducted the analysis of them on the assumption that Plato was completely conscious of their fallacious character. His method of introducing such arguments is, however, not the same as his method in the *Euthydemus*. In that dialogue, the fallacies appeared in the mouth of the two eristic sophists; where explanations were given, either directly or indirectly, they appeared in the mouth of Socrates. In the *Theaetetus*, however, it is Socrates who puts forward the fallacies; the explanations are, for the most part, given by Theaetetus. The shift in method is of course occasioned by the difference in Plato's attitude towards the respondents of Socrates in the two dialogues. Euthydemus and Dionysodorus, as purveyors of a spurious training in virtue, are the legitimate objects of ridicule. Theaetetus, on the other hand, is probably the most promising example of the philosophic mind to be found in the dialogues. Plato's treatment of him therefore differs *toto caelo* from his treatment of the sophists, a fact which comes out particularly clearly in Socrates' speech to Theaetetus at 154Dff.:

Well, if you and I were clever and wise and had found out everything about the mind, we should henceforth spend the rest of our time testing each other out of the fulness of our wisdom, rushing together like sophists in a sophistical combat, battering each other's arguments with counter arguments. But, as it is, since we are ordinary people, we shall wish in the first place to look into the real essence of our thoughts and see whether they harmonize with one another or not at all.

The real difficulty, however, is not to explain why Plato has given Theaetetus a role similar to that which was previously played by Socrates (we are probably right to suppose that this is his way of paying the young student a compliment), but to explain why he has given *Socrates* a role similar to that which was previously played by Euthydemus and Dionysodorus. But the difficulty is only a difficulty if we assume that because Socrates

in the end. Cf. below pp. 50–52. In the *Theaetetus* we may note another close connection between the philosophy of rest and the philosophy of motion at 158Eff. where we find that Socrates ill turns out to be 'wholly other' (κομιδῇ ἕτερος) than Socrates in health. The doctrine of total flux has resulted in an Eleatic use (or rather misuse) of ἕτερος.

is represented as employing eristic arguments, he is represented as employing them in an eristic spirit. If, for example, we compare Socrates' employment of fallacy in the *Theaetetus* passage with its employment by Euthydemus and Dionysodorus in the first scene with Cleinias, it will readily be seen that although the actual arguments are similar, the way in which they are used is quite different:

(1) Socrates is engaged in the testing of a specific definition, knowledge is perception. The sophists, however, have no definite philosophical task in hand; they are doing no more than attempting to substantiate their vague claim to be teachers of virtue.

(2) Socrates wants Theaetetus to grow 163C, whereas the sophists have no interest in Cleinias except to make him look ridiculous.

(3) Socrates shows clearly that the arguments presented are not his own arguments but are those of contestants (*ἀγωνισταί*) who behave *ἀντιλογικῶς* 164C, or of 'a nimble fighter, fighting for pay in the war of words' (*πελταστικὸς ἀνὴρ μισθοφόρος ἐν λόγοις ἐρόμενος* 165D). In the *Euthydemus*, however, there is no doubt but that the sophists are intended to be completely identified with their own arguments; they are in fact precisely the 'nimble fighters' whom Socrates has in mind. (Cf. *ἐν τοῖς λόγοις μάχεσθαι* 272A.)

(4) Socrates handles his arguments in such a way as to bring out the important facts that (*a*) although there is some connection between knowledge and perception, the complete identification of the two would restrict the meaning of 'knowledge' in such a way as not to allow for two additional meanings, i.e., learning and memory, and (*b*) that the 'knowing—not knowing' dichotomy is at the basis of a certain group of paradoxes which can be produced by use of the identification in its extreme form. The sophists in the *Euthydemus* are using similar arguments but with no thought of exposing the mechanics of these arguments—quite the contrary in fact. It is left for Socrates to point out that *μανθάνειν* sometimes means 'learn' and sometimes means 'understand' (278A); if this equivocation were pointed out by Euthydemus or Dionysodorus, the paradoxes they have just set up would be destroyed.

All in all, we may conclude that Socrates' use of fallacy in this section of the *Theaetetus* is as clearly constructive as its use by the sophists in the *Euthydemus* is destructive. In a sense, of course, Socrates has used fallacy destructively, i.e., to destroy the complete equation of knowledge with perception. But he has performed this destructive task in such a way as to show that the definition of knowledge will be incomplete unless it accounts for the positive facts of learning and memory. Furthermore, by the rejection of the definition of knowledge as perception, he has paved the way for the two definitions which Theaetetus will offer later on: that knowledge is *δόξα ἀληθής* 187B, and that it is *δόζα ἀληθὴς μετὰ λόγου* 201D, each of which is an advance over the definition which has gone before. It is clear, then, that fallacy has played an important part in the progress of the dialogue in the direction of a satisfactory definition of knowledge.

In the next chapter we shall again encounter fallacy consciously employed by Plato, but shall find it occurring not in the mouth of Socrates but in that of his opponent Cratylus. In offering a solution to the fallacy, Plato will again be doing constructive work, since he will be lending support to the theory of Forms.

CHAPTER THREE

Cratylus 429B ff.

AT 427D, Socrates has finished the long list of etymologies which, at the request of Hermogenes, he has produced to illustrate the theory of the natural correctness of names. According to Hermogenes, this theory is the one held by Cratylus, but Cratylus himself has not yet entered the conversation. Now, however, Socrates asks Cratylus for his assistance in reconsidering the theory, since he, Socrates, has begun to disbelieve in his own wisdom 428D. Cratylus readily agrees with Socrates that names are given for the purpose of instruction, and that this instruction is an art whose artisans are the law-givers, as Socrates has said before at 388E. But when Socrates wishes to show, by an analogy between naming and painting, that some law-givers do their work better than others, Cratylus objects; he will not concede that one name is better and another worse. He maintains, on the contrary, that all names are correct, all that are really names, that is. Further questioning by Socrates reveals the source of Cratylus' reluctance to admit the existence of names which are false or incorrect:

Soc. How about the name of our friend Hermogenes, which was mentioned a while ago? Shall we say that it is not his name at all, unless he belongs to the race of Hermes, or that it is his name, but is incorrect?

Cra. I think, Socrates, that it is not his name at all; it appears to be his, but is really the name of some one else who possesses the nature that makes the name clear.

Soc. And when anyone says that our friend is Hermogenes, is he not even speaking falsely? For perhaps it is not even possible to say that he is Hermogenes, if he is not.

Cra. What do you mean?

Soc. Do you mean to say that it is impossible to speak falsehood at all? For there are, my dear Cratylus, many who do so, and who have done so in the past.

Cra. Why, Socrates, how could anyone who says that which he says, say that which is not? Is not falsehood saying that which is not?

Soc. Your reasoning is too clever for me at my age, my friend. However, tell me this: Do you think it possible to speak falsehood, but not to say it?

Cra. Neither to speak nor to say it.

Soc. Nor utter it or use it as a form of address? For instance, if some one should meet you in hospitable fashion, should grasp your hand and say, 'Well met, my friend from Athens, son of Smicrion, Hermogenes', would he be saying or speaking or uttering or addressing these words not to you, but to Hermogenes—or to nobody?

Cra. I think, Socrates, the man would be producing sounds without sense.

Soc. Even that reply is welcome; for I can ask whether the words he produced would be true, or false, or partly true and partly false. Even that would suffice.

Cra. I should say that the man in such a case was merely making a noise, going through purposeless motions, as if he were beating a bronze pot. (429B–430A.)

The basis of the position taken by Cratylus is easily recognizable as the Eleatic argument against false speaking with which we are already familiar from the *Euthydemus* 286C. Cratylus' train of thought is apparently as follows: if a name is not the name of the object or person to which it is applied (as, for instance, in the case of Hermogenes, whose names does not express his real nature) then that name is not, or does not exist. Therefore a name either names correctly or is not a name at all.[23] Again as in the *Euthydemus*, the argument depends on the use of

[23] It is possible to see a resemblance between the position taken here by Cratylus about names and that taken by Thrasymachus at *Republic* I 340Cff. to the effect that a ruler who makes a mistake is not really a ruler. Plato, in opposition to both Parmenides and Heraclitus (and Protagoras), is always concerned to allow for error: see, e.g., *Theaetetus* 190E 'if [false opinion] is found not to exist, we shall be forced to admit many absurdities . . .'.

two fallacies: equivocation and *secundum quid.* (A shift in the meaning of 'is' from copulative to existential occurs in conjunction with a shift from the statement that a name is not the name of some particular person or thing, to the statement that it is not absolutely.)

After some further discussion, Cratylus is persuaded to admit that there may perhaps be incorrect names, but his conversion is only temporary and he suffers a number of relapses before the dialogue ends (e.g. 432A, 433C, 438C). The argument against false speaking has influenced him to such an extent that he cannot give it up even when Socrates has demonstrated in the clearest possible way that it is inconsistent to maintain both that all names are correct, and, what Cratylus wishes to say also, that the relation of names to things is one of representation or likeness (ὁμοίωμα). The view that all names are correct is one which Cratylus seems to have arrived at by means of the assumption that a name cannot be in any degree unlike the thing named, or, in other words, it cannot be in any way *not* that thing. But if a name is to be the likeness or image of a thing, it must in some fashion be unlike or not the thing, else it would be an exact copy of that thing. (Socrates illustrates this latter point by the example of the two Cratyluses at 432Bff.) Therefore if Cratylus wishes to retain a theory of naming by representation and likeness (a theory which he much prefers to naming by chance signs 434A), he will need to give up the view that there are no degrees of correctness in names. And he will also, by implication, need to give up the fallacious argument against false speaking which is associated with it.[24]

Now it appears from various passages in the *Cratylus* that Plato is extremely anxious to defend the view that there are

[24] At *Sophist* 240Bff. we see again that there is incompatibility between Eleaticism and the view that images or likenesses have some real status:

Str. That which is like, then, you say does not really exist, if you say it is not true.
Th. But it does exist, in a way. (Ἀλλ' ἔστι γε μήν πως.)
Str. But not truly, you mean.
Th. No, except that it really is a likeness.
Str. Then what we call a likeness, though not really existing, really does exist?
Th. Not-being does seem to have got into some such entanglement with being, and it is very absurd.

A little further on at 241D there is a passage giving a good illustration of the general clash between Eleaticism and the possibility of false speech: the both true and false names.[25] If, for instance, we go back to the beginning of the dialogue, we find Socrates questioning Hermogenes as follows:

Soc. . . . Is there anything which you call speaking the truth and speaking falsehood?
Her. Yes.
Soc. Then there would be true speech and false speech.
Her. Certainly.
Soc. Then that speech which says things as they are is true, and that which says them as they are not is false. (385Bff.)

From this beginning Socrates then proceeds to argue that the parts of true speech are also true, and the parts of false speech false. Names are the smallest parts of speech so that it is possible to utter both true and false names.[26] After establishing this con-

---

Eleatic Stranger tells Theaetetus that he will have to test the theory of his father Parmenides since

unless these statements [that after a fashion not being is and on the other hand in a sense being is not] are either disproved or accepted, no one who speaks about false words or false opinion—whether images or likenesses or imitations or appearances—or about the arts which have to do with them, can ever help being forced to contradict himself and make himself ridiculous.

[25] In the early part of the dialogue, e.g. 385C, Plato usually speaks of names as 'true' (ἀληθές) or 'false' (ψεῦδος), whereas later he tends to say 'correct' (ὀρθός) and 'incorrect' (οὐκ ὀρθός). But the two sets of terms are clearly interchangeable when applied to names (cf. 430D), so that I have felt free to vary the usage in what follows.

[26] As it stands, this argument looks like an example of the fallacy of division, that is, it takes distributively two attributes, truth and falsity, which should be taken only collectively. (See Richard Robinson, 'A Criticism of Plato's *Cratylus*', *Philosophical Review*, LXV, 1956, p. 328.) On the other hand, the premisses needed to establish the conclusion are both present in the text:

385B13, C14 (A) The parts of true and false speech are true and false.
C9 (B) Names are parts of true [and false] speech.
C16 ∴ (C) Names are true and false.

The fact that most people would not accept (A), either because they hold the Aristotelian view (*de Interpretatione* 16a13ff.) or because at *Sophist* 262–264 Plato himself seems to attach truth and falsity to sentences rather than to names (nouns and verbs), is irrelevant to the structure of the argument here in the *Cratylus*. (Incidentally, I think myself that in the *Sophist* passage there is nothing to indicate that Plato, in attributing truth and falsity to sentences, necessarily implies that they should *not* be attributed to names, but this is another matter.) I think, however, that Robinson is right, when he says, *loc. cit.*, that Plato commits the converse fallacy, that of composition, at 431B.

clusion, his next move is to embark on a refutation of Hermogenes' view that names are right by convention and agreement. This he does by showing Hermogenes that the convention theory, if carried to its logical conclusion, implies the relativism of Protagoras. Protagoras' theory, that man is the measure of all things, would involve Hermogenes in two admissions which he is unwilling to make: first, that no men are better than others, and, second, that no men are wiser than others 386B, C. The convention theory would involve Hermogenes also in a species of universal predication theory, attributed by Socrates to Euthydemus ('that all things belong equally to all men at the same time and perpetually') which has similar results ('for on this assumption also some could not be good and others bad if virtue and its opposite were always equally possessed by all' 386D). The point is that the theory of Protagoras and the theory of Euthydemus succeed equally in destroying intellectual and moral distinctions. If these distinctions are destroyed, then no man is better than another and name-giving cannot be a matter for experts, since no experts exist. And name-giving involves no particular wisdom, if all names are equally true. With these implications apparently in mind, Plato argues, on the contrary, that 'the giving of names can hardly be as [Hermogenes] imagine[s] . . . a task for trifling or casual persons' 390D; names are made by the law-giver with the advice of the dialectician. Since Hermogenes can think of no answer to this, Socrates concludes that

Cratylus is right in saying that names belong to things by nature and that not every one is an artisan of names, but only he who keeps in view the name which belongs by nature to each particular thing and is able to embody its forms in the letters and syllables (390D, E).

This is not in fact exactly what Cratylus is reported to have said nor what he will say later on, but the present point at issue is that Hermogenes is confronted with the failure of his theory (the convention theory) and that, furthermore, this failure is connected with Socrates' assertion that there are both true and false names. In other words, the convention theory is associated with the view that all names are true, whereas the natural rightness theory (in its present guise as a theory involving expert knowledge of the nature of the things named) is associated with the view that there are false names as well.[27] The two theories are obviously incompatible, as incompatible, one might say, as are the theories of Protagoras and Euthydemus with the theory of Forms. (The convention theory implies both of the former, as we have seen, and the natural rightness theory, as it is here developed in connection with the notion of the ideal shuttle, certainly appears to imply the latter.) It seems, therefore, as if the existence of false names were an essential element not only in the defeat of the convention theory but in the establishment of the theory which Plato wishes to substitute for it, that is, the theory that the law-giver gives names with his eye on the ideal name, and that his work is judged by the dialectician. This giving and judging of names is obviously work for experts and requires the existence of both wisdom and wise men. And if wisdom exists, so does ignorance, and it is possible for names to be more or less well given. The well-given names are true and the others false. The false names are none the less names, but they, presumably, have been devised in ignorance of, or without attending to, the ideal names.

This entire passage is of great interest, since it shows, in a fairly small compass, the essential points of Plato's quarrel both with the Heracliteans and with the Eleatics. Protagoras, who here, as in the *Theaetetus*, is grouped with those who favor a universal flux, is wrong, since on his view all men are equally wise and good. Euthydemus, who is here as much of an Eleatic monist as he is in his own dialogue (the view attributed to him at *Cratylus* 386D is a fair corollary of the doctrine of universal knowing which he and Dionysodorus maintain at *Euthydemus* 294ff.), is also wrong, since on his view the conclusion is the same: all men are equally wise and good. In the case of Protagoras, the conclusion comes about because each man is assumed to be the judge of his own shifting perceptions; in the case of Euthydemus, the conclusion is the result of a sophistic trick which argues, e.g., from knowledge of some particular thing to knowledge absolute. The two situations are totally different,

[27] This distinction between *two* versions of the theory of natural rightness (one associated with the theory of Forms and the other with the Eleatic position to be taken up by Cratylus 429Bff.) is one not normally made but is, I believe, present in the text and is an important element in my general interpretation of the dialogue. See below pp. 52–55.

but the outcome is the same—no man is better than another and no man is wiser than another. Wisdom and virtue (and their opposites) have, in effect, ceased to exist.

That this is a state of affairs which would be intolerable to Plato need hardly be said; the theory of Forms is inextricably tied to the view that there can be a progress from ignorance to knowledge, and hence that there must be allowance made for the possibility of error. It is also associated with the existence of an intellectual and moral aristocracy. Since the *Cratylus* is concerned with language, Plato has expressed his dissatisfaction with the two opposed positions by arguing not for the existence of folly or of error but for that of false names, but his intention to contrast the theory of Forms with both Heracliteanism and Eleaticism seems to me none the less clear. This contrast is concisely expressed at 386D when Socrates, after ascertaining that Hermogenes is unwilling to accept the views of either Protagoras or Euthydemus, formulates the view which is, in his mind, the necessary alternative to these:

Soc. Then if neither all things belong equally to all men at the same time and perpetually [the position of Euthydemus] nor each thing to each man individually [the position of Protagoras], it is clear that things have some fixed reality of their own, not in relation to us nor caused by us; they do not vary, swaying one way and another in accordance with our fancy, but exist of themselves in relation to their own reality imposed by nature.

When Hermogenes has agreed to this, Socrates then goes on to develop the analogy between the ideal shuttle and the ideal name.

Up to the point at which Socrates actually begins to illustrate the natural rightness theory, then, we have some reason to associate this theory with the theory of Forms. This is particularly clear in the passage 390D, E quoted earlier: Hermogenes and the convention theory have been refuted (partly by means of the analogy with Protagoras and Euthydemus, and partly by means of the passage about the ideal names) and Socrates concludes as a result that

Cratylus is right in saying that names belong to things by nature, and that not every one is an artisan of names, but only he who keeps in view the name which belongs by nature to each particular thing and is able to embody its forms in the letters and syllables.

Before Socrates begins to illustrate the theory, however, we are warned that it will probably not prove satisfactory. Socrates tells Hermogenes 391B that the best way to investigate the theory will be with the help of those who know. Hermogenes could, for instance, ask his brother Callias to tell him what he has learned from Protagoras on this topic—a suggestion Hermogenes naturally rejects since he has already declared himself to be out of sympathy with Protagoras' views 386A. Or he could learn about the correctness of names from Homer. This suggestion Hermogenes for the moment accepts, but we can guess that here as elsewhere in Plato the appeal to Homer is ironic and implies that not much reliance is to be placed on what follows. As the illustrations proceed, we receive further indications that Plato intends the whole performance as a reduction to the absurd (e.g., the references to the inspiration derived from Euthyphro), but Cratylus, whose theory is supposedly being supported and strengthened all this time, expresses no dissatisfaction when Socrates concludes—he says in fact 428C that Socrates' 'oracular utterances [are] much to his mind'. We may therefore suspect that the version of the natural rightness theory which Cratylus would wish to uphold is not one which would be altogether compatible with the theory of Forms. This suspicion is immediately confirmed when we come to the passage cited at the beginning of this chapter, since we find there that Cratylus, while he is willing to agree with Socrates that names are given with a view to instruction, and that this instruction is an art and has its artisans, parts company with Socrates on the matter of the relative correctness of names, maintaining, as we noted earlier, that a name is either correct or not a name at all. Thus he is clearly not a subscriber to the natural rightness theory in the version attributed to him by Socrates at 390D, E (which I have interpreted as a version involving the theory of Forms) since he denies one of its fundamental tenets, 'that not every one is an artisan of names'. On Cratylus' present view, every one who gives a name would be a competent name-giver, and all names would be correct—if the name is incorrect it is simply not a name. Cratylus' reliance on the fallacious argument about false speaking, therefore, upsets the relationship between the natural rightness theory and the theory of Forms. If names are given by the expert who has his eye on the ideal name, then names will in fact possess a natural rightness, but Cratylus, by denying the need for such experts (this is not explicit but is the obvious corollary of his denial of false names), has really also denied the necessity for the vision which supplies the expert with the knowledge which he needs to carry out his linguistic task. Even if he were to admit the existence of the Forms, he would have made it impossible to construct a theory of naming which utilizes knowledge of them, since to say that a name is either a correct name or not a name at all would be equivalent to saying that a name is identical with the ideal name or not a name at all. In other words, he has ruled out the possibility that the name is like the ideal name, or in some degree resembles it, and yet at the same time is other than the ideal name. The elimination of likeness or resemblance results in committing him to a situation in which he cannot really explain how names can convey anything at all about the essential nature of the things named; he cannot do this because he has dispensed with any relationship between them except that of identity. A name which was exactly like the ideal name would tell us no more about the nature of the thing named than a second Cratylus would tell us about the nature of the original Cratylus (432Bff.). And if the name were in any degree different from the ideal name, it could not, on Cratylus' view, tell us anything either, since it would have ceased to exist.[28]

[28] It is of interest to note that an objection to the possibility of incorrect names which is obviously Eleatic in origin comes from the supposed Heraclitean, Cratylus. G. S. Kirk, in an article 'The Problem of Cratylus', *American Journal of Philology*, LXXII, 1951, pp. 225–53, has maintained that 'Plato does *not* depict Cratylus as a convinced Heraclitean' (p. 225) and has noted (p. 230) the Eleatic character of the objection. Kirk's thesis has been attacked by D. J. Allan (in an article with the same title (*A.J.P.*, LXXV, 1954, pp. 271–87)) but Allan's criticism is directed almost wholly towards Kirk's treatment of the Aristotelian evidence concerning the relationship between Plato and the historical Cratylus and seems to me not to face the problem of what actually goes on in the dialogue. (In connection with Allan's paper, H. Cherniss, 'Aristotle, *Metaphysics* 987A32–B7', *A.J.P.*, LXXVI, 1955, pp. 184–86, should also be consulted.) I think myself that the undeniable combination of Heraclitean and Parmenidean elements in the character of the Platonic Cratylus is partly meant as one more illustration of the fact, which Plato sees very clearly, that the two philosophies lead equally to the obliteration of the distinction between truth and falsity. I think Plato means to say also that if one persists in trying to find out the

In order to refute Cratylus' version of the natural rightness theory, then, it is clear that Plato must devise some means of establishing the existence of false names, and this he does through his analogy between naming and painting 430Bff.

Soc. Let us see, Cratylus, if we cannot come to terms somehow. You would agree, would you not, that the name is one thing and the thing of which it is the name is another?

Cra. Yes, I should.

Soc. And you agree that the name is an imitation of the thing named?

Cra. Most assuredly.

Soc. And you agree that paintings also are imitations, though in a different way, of things?

Cra. Yes.

Soc. Well then . . . can both of these imitations, the paintings and the names, be assigned and applied to the things which they imitate, or not?

Cra. They can.

Soc. First, then, consider this question: Can we assign the likeness of the man to the man and that of the woman to the woman, and so forth?

Cra. Certainly.

Soc. And can we conversely attribute that of the man to the woman, and the woman's to the man?

Cra. That is also possible.

Soc. And are these assignments both correct, or only the former?

Cra. The former.

Soc. The assignment in short, which attributes to each that which belongs to it and is like it.

Cra. That is my view.

Soc. To put an end to contentious argument between you and me, since we are friends, let me state my position. I call that kind of assignment in the case of both imitations—paintings and names—correct, and in the case of names not only correct but true; and the other kind, which gives and applies the unlike imitation, I call incorrect and, in the case of names, false.

Cratylus, while willing to concede Socrates' point about incorrect assignment in the case of painting, will not at once

nature of reality by studying names, there is one sense in which it hardly matters what view one takes of the character of these names. The Heraclitean view has, of course, some extra difficulties of its own, but the really important thing is to get things in the right order, i.e., to turn our attention first to τὰ ὄντα, then to language.

concede the point about names—he suggests that the two cases may be different. However, after Socrates again emphasizes the fact that both paintings and names are imitations, he is persuaded to give in 431A.

In this way, by getting Cratylus to admit that there may be names which are assigned incorrectly and are false, Socrates intends 'to put an end to contentious argument' between himself and Cratylus. That is, he intends to get rid of the fallacious argument about false speaking which is doing damage to the natural rightness theory in the version compatible with the theory of Forms.[29]

Cratylus and his theory have thus been dealt with in the same way as Hermogenes and the convention theory. In the opening pages of the dialogue (383A–384E) the two theories are presented as opposed to each other, and so they are in many ways, but there is one important thing they have in common: both assume that all names are true. Therefore the way to refute both theories is to show that, on the contrary, there are false names as well.

The refutation of these two theories of language is, of course, one of the essential tasks accomplished in the *Cratylus*, but it is not the main purpose of the dialogue except in the indirect sense

[29] It is worth noticing that when Plato is in the process of developing the doctrine of recollection in the *Phaedo*, the principle that a painting may be like the object it represents is assumed to hold good 73E. If Cratylus had been present at the discussion, we might have expected the objection, how can the painting exist if it is not the thing painted? In the *Phaedo*, however, Simmias and Cebes are both in sympathy with the theory of Forms, and this type of objection is not raised. The doctrine of recollection involves a gradual recovery of knowledge, so that at any point in the process it is difficult to say whether the person recollecting either knows or does not know a particular thing. We have already seen from the *Theaetetus* that the Eleatic insistence on a choice between knowing or not knowing is incompatible with the fact of memory. We can see a similar incompatibility if we compare the remarks on painting in the *Cratylus* with those in the *Phaedo*: the principle of likeness in painting is criticized in the *Cratylus* for reasons which stem ultimately from the knowing—not knowing dichotomy and the denial of false speaking; it is accepted in the *Phaedo* in a context in which there is also acceptance both of the fact of memory and of its constant companion, the theory of Forms. (It should also be noted that when the doctrine of recollection is introduced in the *Meno*, it appears there as the direct answer to an Eleatic difficulty about the possibility of knowledge 80D. See below pp. 84–86.)

in which its accomplishment lends support to the theory of Forms. The primary concern of the dialogue, as I see it, is to support the theory by answering the following three questions: (1) Can the nature of reality be discovered through the study of language, to which the answer is quite definitely, no; (2) How then can it be discovered, to which the answer is, through direct knowledge of *τὰ ὄντα*; and (3) What do we find out about *τὰ ὄντα* by this method, to which the answer is, we find out that there are really existing things such as absolute beauty, etc. These three questions are treated with varying degrees of length and answered with varying degrees of definiteness, but that they comprise the substance of the dialogue seems to me clear.

The arguments against the ability of the study of language to provide us with reliable information about the nature of reality are chiefly two. The first is that there seems to be no consistent view of reality to be gained from the study of language since it is possible to construct etymologies of rest which are just as convincing as the etymologies of motion (437Aff.). We can guess, of course, from Plato's remarks at 411B and 439C about the philosophers who make themselves dizzy with turning around and therefore suppose that reality turns too, that he would, if forced to make a choice, prefer the etymologies of rest, but that he regards the whole etymological method as wrong is obvious from the second and more fundamental objection to it which Socrates puts forward at 438Aff.:

Soc. . . . A little while ago . . . you said he who gave names must have known the things to which he gave them. Do you still hold that opinion, or not?

Cra. I do.

Soc. And do you say that he who gave the first names also knew the things which he named?

Cra. Yes, he knew them.

Soc. But from what names had he learned or discovered the things, if the first names had not yet been given, and if we declare that it is impossible to learn or discover things except by learning or ourselves discovering the names?

Cra. I think there is something in what you say, Socrates.

Soc. How can we assert that they gave names or were law-givers with knowledge before any names whatsoever had been given, and before they knew any names, if things cannot be learned except through names?

Cra. I think the truest theory of the matter, Socrates, is that the power which gave the first names to things is more than human, and therefore the names must necessarily be correct.

Soc. Then, in your opinion, he who gave the names, though he was a spirit or a god, would have given us names which made him contradict himself? Or do you think there is no sense in what we were saying just now?

Cra. But, Socrates, those that make up one of the two classes are not really names.

However, since Socrates has previously shown that some names can be derived from motion and some from rest, Cratylus has no way of deciding which of these two opposed sets are to be the true names. He is therefore constrained to agree when Socrates goes on to say 438E:

> . . . it is plain that we must look for something else, not names, which shall show us which of these two kinds are the true names, which of them, that is, show the truth of things.

And since they have also agreed 'that names which are rightly given are like the things named and are images of them', he is also constrained to agree with Socrates' conclusion 439Aff.:

> Then if it be really true that things can be learned either through names or through themselves, which would be the better and surer way of learning? To learn from the image whether it is itself a good imitation and also to learn the truth which it imitates, or to learn from the truth both the truth itself and whether the image is properly made?
>
> *Cra.* I think it is certainly better to learn from the truth.

In this passage Plato has finally followed out to their logical conclusion certain objections which appeared earlier during the long discussion of the various etymologies. At 409D, for instance, Socrates was asked by Hermogenes to explain the word *πῦρ*. 'Either,' replies Socrates, 'the muse of Euthyphro has deserted me or this is a very difficult word.' However, he has a contrivance (*μηχανή*) to deal with such cases; he attributes them to foreign sources rather than trying to explain them in terms of other Greek words. All that he has done, however, is to put an arbitrary end to the infinite series of questions which inevitably arises when one attempts to explain names by means of other names, by shifting the burden to another language. Socrates goes into the topic of the contrivance and its relation to an infinite series of explanations much more thoroughly at 421Dff. Here he suggests what we have probably already suspected, that the device of foreign origin is not the real solution of the difficulty, since an ancient Greek word might be identical with a modern foreign one. (In this case the explanations would necessarily continue.) Furthermore, he points out that if the person giving the explanations is pressed hard enough, he will eventually give up. The contrivance was, of course, a way of not giving up, but since this has failed, something else will have to be devised. What Socrates presently suggests is to see whether it might not be possible to carry the investigation back a stage further by making the ultimate elements letters instead of names. We can guess in advance that this will not work, since Socrates is still proceeding on the assumption (to be rejected 438E) that the study of language in some form will finally succeed in showing the nature of the things named. Even if this method should bring us to the point at which the ultimate elements imitate reality, the problem is still not completely solved, since if language is taken as genuinely imitative, it would seem as if the correct way to name, say, a cock, would be by crowing, not by uttering the word 'cock'. As Socrates says at 425D,

It will, I imagine, seem ridiculous that things are made manifest through imitation in letters and syllables; nevertheless it cannot be otherwise. For there is no better theory upon which we can base the truth of the earliest names.

To use the theory of foreign origin or to say that the god gave these names (as we have just heard Cratylus suggest at 438C) 'are merely clever evasions on the part of those who refuse to offer any rational theory of the correctness of the earliest names' (426A).

Socrates is of course right to insist that if the investigation of reality is to be conducted by means of the study of language, it is essential to explain correctly the linguistic elements which are the direct imitators of reality, since 'if anyone is, no matter why, ignorant of the correctness of the earliest names, he cannot know about that of the later'. As in a series of deductions in geometry, the initial error may be small, but it will permeate the whole series. (Socrates uses this illustration later at 436D.) What Socrates does not declare openly here (although he gives us a hint by saying 426B 'I think my notions about the earliest names are quite outrageous and ridiculous') is that the whole etymological procedure is unsatisfactory. We do not get his clear objection to it until 438A, B (quoted above), i.e., that the giver of the earliest names could never have given these names correctly if he were solely dependent on other names in order to learn the true nature of things, since at that time, no names as yet existed.

As soon as this objection has been made, it becomes possible for Socrates to begin looking for 'something else, not names' (438D) as a criterion for judging which names are correct. It appears, after all, 'that things may be learned without names' and that the way these things are learned is 'through each other, if they are akin, and through themselves'. The names which are rightly given are like the things named and are images of them, but to learn whether these images are properly made, it will be best to learn the truth (i.e., *τὰ ὄντα*) first, rather than (as has been the attempted method throughout the dialogue) to attempt to learn the images first. In other words, the proper order has been restored to the study of language; reality is to be apprehended first, after which it will be possible to judge whether a given name portrays this reality correctly.

It now becomes apposite for Socrates to hazard some views as to the nature of this reality. So long as he continued to investigate the claim of language to be an index to *τὰ ὄντα*, what he himself thought about *τὰ ὄντα* was irrelevant. There have of course been a number of hints that Socrates prefers rest to motion, but he cannot embark on an exposition of this view until Cratylus has agreed that the correct way to learn the truth (and also to learn what names are properly made images of this truth) is from the truth itself and not from language. As soon as Cratylus has made this admission, however, the time has come for Socrates to point out the difficulties in the view that reality is in total flux. At 439C he speaks as follows:

> Suppose it should prove that although those who gave the names gave them in the belief that all things are in motion and flux . . . still in reality this is not the case, and the name-givers them-

selves, having fallen into a kind of vortex, are whirled about, dragging us along with them. Consider, my worthy Cratylus, a question about which I often dream. Shall we assert that there is any absolute beauty or good or any other absolute existence, or not?

Cra. I think there is, Socrates.

Absolute beauty, Socrates continues, is always such as it is. If it were always passing away, we could never say that it was this or that, since it would have changed in the very instant of our speaking. A thing which is never in the same state cannot really be anything, nor, in fact, can it ever be known by anyone. Knowledge requires that that which knows and that which is known must always be—a set of conditions which bears no relation to the condition of flux or motion. Of course, Socrates concludes, the doctrine of Heraclitus and others may be true, but we are surely left in no doubt as to his opinion of this doctrine. In any case, 'no man of sense can put himself and his soul under the control of names, and trust in names and their makers to the point of affirming that he knows anything'. Socrates is content to have discredited the study of language as a means of apprehending *τὰ ὄντα*; what this reality is like, he insists on less forcibly, but if we have been alert throughout the dialogue, we can hardly fail to note that the theory of Forms has been advocated all along the line.

This advocacy of the theory of Forms seems to me, as I have said, to be the primary purpose of the dialogue,[30] and the refutation of the two theories of language a subsidiary and auxiliary one. Nevertheless, Plato is certainly not intending to

[30] If the theory of Forms really is the dominant theme of the dialogue, it may appear strange that so much space has been given to etymology. It is clear not only from the *Cratylus* but from passages in other dialogues that Plato was interested in etymology for its own sake (cf. *Protagoras* 361D; *Phaedo* 81C; *Phaedrus* 237A; *Theaetetus* 194C; *Sophist* 221B), but he certainly did not expect to gain true knowledge by this means. As far as the structure of the dialogue is concerned, I am quite willing to agree with Méridier when he writes:

On convient en général que la composition du *Cratyle* n'offre pas la belle ordonnance ni l'équilibre si sensibles dans d'autres dialogues platoniciens. On s'est montré surtout choqué du développement disproportionné donné à la partie étymologique. Platon croyait avoir ses raisons; mais il est certain que du point de vue artistique l'économie de l'oeuvre en a souffert. (*Introduction* to the Budé edition, Paris, 1931, p. 33.)

discredit the use of language altogether. Some things, he says, can be learned without language, but this sort of learning is presumably restricted to the name-maker who has a vision of the Forms and thus constructs the earliest names correctly, and to the dialectician who is the judge of the name-maker's work.

What Plato's own theory of language would be like, we can guess with, I think, a fair degree of accuracy:

(1) It is based on knowledge of the Forms, and this knowledge is the source of any judgement as to whether a name is correctly or incorrectly made. The error of the convention theory and of Cratylus' version of the natural rightness theory was that both theories, in addition to denying the existence of false names, put things in the wrong order; they attempted to discover the nature of reality through the analysis of names (in which case they had no way of telling whether the names themselves were correct) instead of apprehending reality first (in which case the criteria for correctness naturally follow).

(2) It will insist that there are degrees of correctness in names, or, in other words, that there are false names as well as true. This, of course, is closely tied to point (1) since Plato's view that names, to be correct, must be given with knowledge of the Forms, implies, as Socrates says at 390D, that name-giving is a task for experts. As we have already seen, both the convention theory and the natural rightness theory militated against the need for expertness in name-giving (and therefore against any need for knowledge of the Forms) by arguing, each in its own distinctive way, that all names were true.

(3) It will involve some theory of imitation. Socrates says at 435C:

> I myself prefer the theory that names are, so far as is possible, like the things named; but really this attractive force of likeness is, as Hermogenes says, a poor thing, and we are compelled to employ, in addition, this common-place expedient, convention, to establish the correctness of names.

Socrates is not meaning to say that he admires the natural rightness theory in the version propounded by Cratylus (since this version denies the existence of false names) but is indicating that, while the best situation would be one in which a name would imitate the essence of the thing named precisely, this is not possible. (If the name did this, it would be the thing, in the sense of being indistinguishable from it.) Therefore some conventional elements must be admitted—in the case of numbers, in fact, the names will have to be completely conventional 435B. Plato sees a great many difficulties in a theory of imitation. The Eleatic objection to images (that they are not the thing and therefore do not exist) is an artificial one, and this he can dispose of readily. But the exact way in which a name can be sufficiently imitative of a thing to convey its essence remains mysterious. We do not name a cock by crowing or a horse by galloping, and even if we did so, this method would not take account of such things as numbers. Again, naming is something like painting, but this analogy is not exact. The most that can be said is (423E) that '. . . if anyone could imitate the essential nature of each thing by means of letters and syllables, he would show what each thing really is . . .'. The precise nature of this imitation Plato (I think quite rightly) does not attempt to explain.

What Plato's theory of language might be like beyond these three points it is probably not possible to say. (Nor, it should be said, am I concerned with whether the theory is an adequate theory by contemporary standards.) As it is, however, the three points I have suggested provide us with its basic outlines: names are given with relation to the Forms, some are better than others (since some name-givers are presumably more skillful than others), and they in some fashion imitate the Forms. These points are all closely interrelated: e.g., the varying degree of correctness in names is related both to the degree to which they imitate the Forms successfully and to the necessity for skill in name-giving. Altogether they form a basic complex of doctrines into which it is possible to fit elements as diverse as Plato's objections to the relativism of Protagoras (385Eff.), or his observation that a name may contain many inappropriate elements and yet be a correct name 432E, or his remarks about the effect of a small initial error on a chain of geometrical deductions 436D.

Into this complex of doctrines it is clear that we may also fit Plato's exposition and treatment of the Eleatic fallacy against false speaking with which this chapter began. It was on the basis of this fallacy that Cratylus wished to maintain that all names are correct, since a name which is incorrect is simply not a name at all. This is of course destructive both of the requirement of skill in the giving of names and of the possibility of any theory that names are images or imitations of Forms. The fallacy of false speaking, therefore, appears in this dialogue as a real threat both to the theory of Forms and to a theory of language based upon the theory of Forms. Plato has introduced the fallacy, with his usual dramatic skill, to illustrate once more the fundamental incompatibility of Eleatic monism with his own philosophical scheme.

CHAPTER FOUR

The *Hippias Minor*

THE passages from the *Theaetetus* and the *Cratylus* which have been studied in the two preceding chapters were chosen with a view to providing really close parallels to arguments in the *Euthydemus* since, as was stated in the introduction, this seemed to be a good way of establishing a few strong cases for the thesis of conscious use of fallacy in Plato. In the present chapter, I propose to diverge from this procedure slightly, and to go to a dialogue which, although it employs one of the fallacies prominent in the *Euthydemus*, equivocation, does not make use of exactly the same terms. (That is, the equivocations in the *Euthydemus* were primarily on 'learn' and 'is'; here they will be on 'good' and 'voluntary'.) It might as a result be argued that the use of fallacy in the *Hippias Minor* is not necessarily conscious, since Plato might well be aware of the ambiguity of some terms but not of others. On the other hand, there appears to be independent evidence in the *Hippias Minor* that the equivocations here are in fact consciously used, so that in this case there is no need to invoke the principle of a precise parallel with arguments in the *Euthydemus*. The *Hippias Minor* may therefore be regarded as a dialogue which extends the range of equivocations which Plato seems to have employed consciously, and also as strengthening the case for the more general thesis of conscious use. Certainly we can again see that the study of fallacy leads us directly to a number of the really critical philosophical problems in Plato's thought.

The dialogue consists of a single argument (with variations) issuing in the paradoxical conclusion that it is the good man who errs voluntarily. Socrates is not himself satisfied with this conclusion, but, as he says in his final speech 376C, it is 'the inevitable result of our argument'. We may therefore find it profitable to examine this argument with a view to finding out just how Socrates has maneuvered himself (and Hippias) into a position which, if we know anything about his convictions, seems to have been completely contrary to the one he held. We may then inquire what Plato intended to accomplish by the production of the paradox.

At the beginning of the dialogue, Hippias has just finished a display on the subject of Homer. Socrates, although silent at first, does have one or two points on which he would like to question Hippias; in particular, he would like to know his opinion of Achilles and Odysseus: which is the better man, and in what respect? Hippias answers 364C that of those who went to Troy, Homer made Achilles the bravest, Nestor the wisest, and Odysseus the wiliest. This answer does not meet the question squarely, since Hippias (*a*) has given Homer's opinion, not his own, (*b*) has not said, except by implication, which man either he or Homer considers better, and (*c*) has confused the issue by mentioning Nestor as well. Furthermore, Hippias' answer has not provided Socrates with quite the material he needs in order to bring off his initial refutation of the sophist at 369B.

If we look ahead to this refutation we see that Socrates there points out the incompatibility in two statements made by Hippias: (1) that Achilles is true, and Odysseus false (and wily), and (2) that the true man and the false man are in fact the same. In order to arrive at this incompatibility it is clear that Socrates will need first of all to set up a strong opposition between Achilles and Odysseus, and, second, to contrive some means of destroying the opposition he has just set up. Let us see how he goes about it.

Socrates first quietly drops Nestor, whom he does not need, and proceeds to concentrate on the contrast between Achilles and Odysseus. Hippias has characterized Achilles as brave and Odysseus as wily, but this does not oppose the two sufficiently for Socrates' purposes. He therefore pretends not to have understood what Hippias meant by calling Odysseus wily, and asks whether Homer did not make Achilles wily as well. He now gets the desired response, since Hippias replies 365B that Homer made Achilles true and simple and Odysseus wily and false. Socrates then proceeds to make the opposition as complete as possible by saying 365C 'Homer . . . thought that a true man was one man and a false man another, but not the same.' Since, furthermore, his purpose is to refute not Homer but Hippias, he next makes sure that Hippias agrees with this statement. Having done this, he dispenses with Homer altogether (on his customary grounds that it is impossible to ask an absent writer what he means[31]) and addresses himself exclusively to Hippias.

Socrates has now established one of the two statements necessary to produce the refutation at 369B; he has induced Hippias to assert that, in his opinion, Achilles is true and Odysseus false, and that the true man and the false are in no wise the same.

Socrates' next step is to show that while Hippias has asserted Achilles and Odysseus to be totally different, he can also be made to assert that they are completely the same. This Socrates does by attributing to the false man a number of properties which are by nature ambiguous and which therefore cut across the opposition between the false man and the true. The first of these is power: this is an ambiguous attribute since to say simply that a man is powerful does not specify whether he is employing his power for good or for evil. Socrates has masked the ambiguity by the phrasing of his question; he says 365D

Do you say that the false are, like the sick, without power to do anything, or that they have power to do something?

Hippias falls into the trap by giving the natural answer: 'I say that they have great power to do many things and especially to deceive people.' He is clear in his own mind that by admitting power as an attribute of the false man, he has admitted only power for evil. Socrates, however, retains the attribute 'powerful' by itself without any of its specifications. He then goes on to attach another ambiguous attribute to the false man, that of wiliness, which has been associated with Odysseus from the beginning. The usefulness of this move becomes immediately

[31] Cf. *Protagoras* 347E.

clear, since although the attribute 'wiliness' connotes the shiftiness of the false man, it also suggests the intellectual ability which enables such a man to accomplish his designs. Socrates brings this out at 365E, asking whether the false are 'wily and deceivers by reason of simplicity and folly, or by reason of shrewdness and a sort of intelligence?' Here again, Hippias, in choosing the latter alternative, probably has in mind a misused intelligence, but Socrates prefers to retain the attribute in its unspecified form, saying 365E 'they are intelligent then, as it seems'. After adding 'knowing' and 'wise' to the list, he reminds Hippias 366A that 'the true and false are different and complete opposites of one another' (*ἐναντιωτάτους ἀλλήλοις*) and that Hippias has now placed the false among the powerful and the wise. The sphere in which the false are wise and powerful is that of the uttering of falsehoods, Socrates goes on; a man who lacks this power would not be false. At this point Socrates is more or less confirming Hippias in his view that to be powerful and wise means to be powerful and wise for evil, but we shall see that he soon drops this line of thought, and begins instead to develop these two notions in the direction of their capacity for good. In order not to make the transition too abrupt, he begins with an example which is morally indifferent: a man has power who is able to do what he wishes when he wishes, e.g., he can write another man's name if and when he wishes 366C. The next step is to carry the ideas of power and wisdom into the field of intellectual skills, e.g., arithmetical calculations. Hippias' admission that he is wise and powerful in these matters does not immediately lead to any moral judgment, but Socrates takes care to lay the groundwork for this addition, since he induces Hippias to say 366D that if he is most powerful and wisest in these matters, he will also be the *best*. (In other words, he will presently change the meaning of 'good', here 'good at', or 'skilled', to 'morally good'. See below 367C.) At the moment Socrates simply relies upon the latent feelings of approval aroused by the word 'best' to ease the progress from power to tell falsehoods to power to tell the truth. This is the crucial step, because he has now definitely crossed the boundary between truth and falsity; he has previously associated power only with the false, so that if true is really the opposite of false, the true should be the powerless, not the powerful. Socrates now goes on to make his position stronger by dwelling on the fact that the powers for truth and falsity in calculation are to be found in the same man. Furthermore, this man who 'has most power to speak both falsehood and truth about calculations . . . is the one who is good in respect to them'. And no one 'becomes false in respect to calculation other than the good man', who is also powerful and true. The man who is ignorant of calculations, on the other hand, will be bad in the sense of lacking the skill to tell a falsehood when he wishes; he will in fact sometimes accidentally tell the truth through his badness at the subject.[32]

Socrates is now in a position to present Hippias with a preliminary version of the refutation which he will conclude at 369B. (As a matter of fact the refutation is complete now, but Socrates refrains from calling Hippias' attention to the original statement about Achilles and Odysseus until he has brought forward further examples.) He says 367C:

> You see then, that the same man is both false and true in respect to these matters, and the true is in no wise better than the false? For he is indeed the same man, and the two are not utter opposites, as you thought just now.

Hippias answers, 'Apparently not, at least not in this field.' Socrates then repeats the same argument for geometry and astronomy. No really new points emerge with the addition of these two new examples, but some of those implied in the previous one come out more clearly. The bad geometer is mentioned as well as the good, and it is pointed out that he will be powerless to speak falsehood. (Thus the opposition between good and bad is being more explicitly allied with that between true and false, and both oppositions will be destroyed together.) We see also that the man who lacks power in astronomy is characterized not only as being unable to speak falsehood but as being ignorant. Badness, powerlessness, and ignorance are all being separated off from the man who is able to speak both

[32] The notion of power is similarly ambiguous in the *Hippias Major* 295Eff.: *Δύναμις μὲν ἄρα καλόν, ἀδυναμία δὲ αἰσχρόν;* asks Socrates, thus identifying *δύναμις* with the beautiful and the useful. But at 296C it turns out that power and useful things are sometimes useful for the accomplishment of something bad, in which case one would hardly wish to call them beautiful. (The possibility that the beautiful might be powerful for good 296D is investigated next, but this fails on other grounds.)

truth and falsehood, i.e., who is both false and true. Hippias will not, Socrates concludes, be able to find a single art, even of those many in which he himself is expert, in which the same situation does not prevail: the man who is best at truth in any art will also be best at falsehood. Hippias cannot give any exception to this rule, so he is now forced to accept the consequences: that what he has just now admitted about the identity of the true and the false man is directly contradictory to what he said earlier, in the case of Achilles and Odysseus, about their being different. (The inference to be drawn from the identity of the two is that neither is better than the other; thus the original question about the relative merits of Achilles and Odysseus remains unanswered.)

Hippias attempts to shy away from the refutation by complaining that Socrates is only talking about details and not paying attention to the whole subject. He himself wants to return to Homer and to prove by quotations that Achilles really is better than Odysseus, since one tells falsehoods and the other does not. He suggests that both he and Socrates should make speeches on this subject and should let the listeners decide which of them speaks better. Socrates is now willing to make at least a pretence of meeting Hippias on this ground, since he has succeeded in bringing off one refutation and can afford to humor his opponent briefly while he prepares the way for the next. He will show, by quotations of his own, that Achilles is just as wily and false as Odysseus. This proposal produces from Hippias the response which provides Socrates with the material for the refutation occupying the remainder of the dialogue.

Hippias, while he has to admit that Achilles does tell falsehoods, still considers him the better man, since what falsehoods he tells he tells against his will; Odysseus, on the other hand, tells his voluntarily and by design. If Hippias really thinks this, Socrates says, then he should, to be consistent, consider not Achilles but Odysseus the better man, since they agreed earlier that those who uttered falsehoods voluntarily were better than those who did so involuntarily. Hippias is surprised at this, since Socrates seems to be upsetting the common-sense views of forgiveness and punishment. Socrates admits that he is all astray about these matters—sometimes he agrees with Hippias, but at the moment a paroxysm of ignorance has seized him, and those who do evil voluntarily seem to be better than those who do it involuntarily. 'And,' he says, 'I lay the blame for my present condition on the previous argument' (372E).

This is the first really open hint in the dialogue that the conclusion which has been reached at the end of the first refutation is no more acceptable to Socrates than it was to Hippias. Socrates is, however, perfectly clear about the cause of this conclusion; it is the argument which is to blame. Since he has constructed the argument himself, he has certainly not placed himself in this predicament by inadvertence. It seems then as if it must be at least part of Plato's purpose to call attention to the way in which the argument has been constructed. We shall see further reason to think this as the dialogue proceeds.

Next is a brief interlude during which Hippias tries to back out of the discussion on the ground that Socrates always makes confusion and trouble in arguments (373Bff.) Socrates, however, is still eager to investigate the matter they have just mentioned (which are better, those who err voluntarily or those who do so involuntarily), and Hippias is finally persuaded to go on answering questions.

Socrates no longer continues to remind Hippias that the proposition they are now considering is the one that resulted from the first argument, but we shall see that the new argument repeats one part of the first one all over again. The terminology is different ('voluntary' takes the place of 'powerful') and the illustrations are broader and are presented in a way which brings out their moral connotations, but the examples leading to the conclusion that the good man errs voluntarily are constructed on the same lines as the example of, for instance, the calculator who has the power to make mistakes.

The shift in the character of the examples may be seen at once. Socrates has previously restricted himself to intellectual error. Now he begins to speak of bodily actions, a sphere in which it is more natural to think of mistakes as a disgrace. The man who runs slowly, for instance, does a disgraceful act. If he does this voluntarily, however, he is better (and again, of course, 'better' means more skillful) than the man who does it involuntarily, that is, than the man who runs slowly because he lacks the power to run any faster. Hippias, as in the earlier discussion, admits that Socrates is right in this particular case, but he does not immediately concede the general principle. It is thus necessary for Socrates to continue the illustrations at some length, moving from bodily actions to the senses and from there to possessions of various sorts, some attached to the body (such as the feet) and some separate from it (such as various kinds of instruments, both animate and inanimate). The fact that this argument is the same as the first shows clearly when the illustrations swing round again to the intellect. We have (e.g., at 375B) the example of the archer: it is better to possess the mind which voluntarily misses the mark than the one which does so involuntarily. Socrates repeats the point in the case of medicine and of music; in these fields it is again better to possess a mind which errs voluntarily rather than the reverse. Hippias, however, still rebels against the universal conclusion; he says at 375D:

But it would be a terrible thing, Socrates, if those who do wrong voluntarily are to be better than those who do so involuntarily.

Socrates' answer is again to remind him of the source of this view, saying, 'But surely they appear to be so, from what has been said' (ἐκ τῶν εἰρημένων). 'Not to me', replies Hippias.

Socrates then gives one further example, the crucial instance of justice 375D, E. This, he suggests, is either a kind of knowledge or a kind of power or both. If it is a power, then the more powerful soul is the more just, since they have already agreed that the powerful soul is better than the one which lacks power. And if it is a kind of knowledge, the wiser soul is more just than the one which is ignorant. This more powerful and wiser soul is not only better but more able to do both good and disgraceful acts in every sphere.[33] When it does disgraceful acts, it does them voluntarily by reason of power and art, either or both of which are attributes of justice. The doing of disgraceful or evil acts is the same as doing injustice, so that when the powerful and better soul does these acts it will be doing injustice voluntarily.

[33] Cf. the idea put forward in *Republic* I 334A that the just man is the best thief. (Here there is ambiguity in the adjective δεινός; but see also the analysis of D. J. Allan in the notes to his edition of *Republic* I, London, 1953, pp. 89–90.)

This powerful and better soul is the good soul and is the property of the good man. Socrates thus concludes that

> he who voluntarily errs and does disgraceful and unjust acts, Hippias, if there be such a man (*εἴπερ τίς ἐστιν οὗτος*), would be no other than the good man. (376B.)

Hippias is still unable to agree with Socrates, nor, in fact, is Socrates able to agree with himself; again this situation is the 'inevitable result of [the] argument' (376C). Socrates wanders up and down on this matter and is still doing so as the dialogue ends.

If we take Socrates' hint and examine the argument, we can detect the chief sources of error as before. In the first place, the strong type of opposition which occurred earlier between the true man and the false is here continued, but is replaced by a comparable opposition between the good man and the bad. (As a result the ethical connotations are wider.) It continues also to be the case that other oppositions follow, as, for instance, the all-important one between the voluntary and the involuntary. (The voluntary takes the place held by the powerful in the earlier sequence.) When particular illustrations are given, other oppositions are used which are appropriate to the matter in hand, e.g., running quickly vs. running slowly, and so forth. In fact the whole notion of conducting the argument in terms of exhaustive alternatives is simply taken for granted, and Socrates makes no further attempt to call our attention to the procedure. (Earlier, at 366A, he took pains to point out that the true and the false were different and in no way the same.)

Now these oppositions are not themselves arrived at by fallacious means, nor is it in itself a fallacious procedure to employ them in an argument. (In the *Theaetetus* passage, the knowing—not knowing dichotomy was set up by the use of *secundum quid*, but here this is not the case.) Hippias claimed almost from the beginning that Achilles was true and Odysseus false, and appeared to mean, with a little prompting from Socrates, that the two were completely opposed in this. Furthermore, if such terms are employed in a consistently absolute sense throughout the argument, there is no more deception in the conclusion than there was in the premises. Or, to put it another way, if the opposition between true and false or between good and bad is felt to be a misrepresentation of reality (on the ground, say, that no absolutely true or false or good or bad man exists), then the misrepresentation permeates the entire sequence; it is not something which emerges to confront the answerer unexpectedly in the conclusion. Therefore any objection to the opposition between e.g. the true and the false should be made straightway as soon as this opposition appears in the premises. On the other hand, and this is the important point, it may also be said that these oppositions are essential to the success of the ultimate refutations of Hippias (and in fact to any refutation) and so might be said to be one of the sources of error. This point is made clear by Aristotle, who says (*On Sophistical Refutations* 168a37) 'refutation is a proof of the contradictory' (ὁ γὰρ ἔλεγχος συλλογισμὸς ἀντιφάσεως). In other words, if Socrates had attempted to work with the statements 'Achilles is somewhat true', or 'Odysseus is more or less false', he would not have been able to bring off a complete refutation, since refutation is effected by asserting the exact opposite of a statement, not its partial opposite. Thus there is, one might say, a certain static quality necessarily to be found in formal argument, and this static quality is by its very nature incompatible with the correct representation of things in flux. If however the necessity and extent of this misrepresentation be understood, no deception need result. Formal logic can never cope with becoming in all its phases, and this fact simply has to be accepted and faced.

If we move on to consider the argument further, we shall see that the word 'good' is clearly used equivocally. Socrates is working towards the establishment of the paradox that it is the good man who errs voluntarily. The paradox is a paradox because voluntary error conflicts with virtue in the public mind. (Hippias has pointed out that the laws are constructed on the opposite assumption; it is the voluntary evil-doer who is punished, and he is presumably regarded as evil, not good, in the eyes of the law.) 'Good' is certainly intended to mean 'morally good' in this conclusion. When the illustrations begin at 373C, however, with the example of the runner, it is equally certain that 'good' means 'good at' or 'skilled'. The way is paved for the shift in meaning almost immediately, when Socrates says at 373E, 'Then he who runs badly performs a bad and disgraceful (αἰσχρόν) act in a race'. He continues to associate 'disgraceful' with 'bad' at intervals as the illustrations proceed, saying for instance at 375C that in connection with all the arts and sciences the mind is better 'which voluntarily does bad and disgraceful things and commits errors'. By the time the example of justice is reached at 375D, Hippias is well prepared to agree with Socrates when he associates the good man with the performance of disgraceful acts, although he is still unwilling, in spite of the argument, to allow that such acts are done voluntarily. But the good man is not, of course, strictly analogous to the good runner or the good musician. The good runner can be skillful at running without necessarily being morally virtuous, but the good man *must* be virtuous since no other activity than the activity of being a man has been specified for him to be skillful *at*. In other words, in the absence of any other action, he can only be described as good because he is a good human being, and this sort of goodness implies virtue rather than skill.

The term 'voluntary' is also equivocal.[34] This is not altogether obvious when the term is associated with good acts, but becomes so when it is also associated with disgraceful acts or mistakes. The paradoxical conclusion to the present argument is 'the good man errs voluntarily', a statement which is similar to 'the good calculator has the power to make mistakes' in the first part of the dialogue. The voluntary means primarily 'what is in

[34] Here I have followed Taylor's lead (*Plato: the Man and his Work*, p. 37) and have consulted Proclus. The relevant passage is

πλὴν ὅτι κατ' αὐτὸν οὐκ ἔστι τὸ ἑκούσιον καὶ τὸ ἐφ' ἡμῖν ταὐτόν. ἀλλὰ τὸ μὲν ἑκούσιον ἐν μόνοις ἐστὶν τοῖς ἀγαθοῖς, εἴπερ ἀκούσιος καὶ ἀβούλητος ὁ τῶν κακῶν βίος· ἐφ' ἡμῖν δὲ καὶ ἀπὸ τῶν ἡμετέρων αἱρέσεων καὶ τὰ ἁμαρτήματα. καὶ γὰρ ταῦτα αἱρούμεθα καὶ ἑλόμενοι πράττομεν, ἀλλὰ δι' ἄγνοιαν. εἰ οὖν τὰ ἁμαρτήματα ἀκούσια μέν, ἐν ταῖς αἱρέσεσιν δὲ ἡμῶν ἐστι καὶ ταῦτα, πάντα μὲν τὰ ἑκούσια καὶ αἱρετὰ δήπουθέν ἐστιν, οὐ πάντα δὲ τὰ αἱρετὰ ἑκούσια.

(*In Platonis Rem Publicam Commentarii*, ed. Kroll, Leipzig 1899, 2 vols., vol. 2, p. 355.)

'However he says that the voluntary and the "what is in our power" are not the same. For the voluntary is concerned with goods only (since a life of evils is involuntary and unwished for) but mistakes are also in our power and arise from our own choices. We indeed choose these, and, having chosen them, do them, but through ignorance. If, then, mistakes are involuntary, and even these are among our choices, all voluntary things, I suppose, are also chosen, but not all things chosen are voluntary." (My translation.)

our power', but it has also the connotation of what we wish or desire. Now in the example about the calculator there was no suggestion that *qua* calculator he could possibly wish to make a mistake; a situation in which he might be called upon to do so would be highly artificial and hypothetical only. Again in the case of the wrestler: presumably he has the power to allow himself to be thrown, but *qua* wrestler he cannot really desire to be thrown because he would lose the match. If we carry this notion over into the case of the good man, we can see that the situation is again hypothetical: the good man has the power to err, but cannot, by the very description of his nature, have the desire to do so. (How can a good man desire what is bad any more than can a fast runner desire to run slowly?) Plato gives us the key to all this when he says in Socrates' next to last speech, 'if there be any such man'; we are to infer that there is none.[35] If the good man performs only good acts, it is correct to describe him as performing them voluntarily since there is no conflict between what is in his power and what he desires. (The former classification would simply be broader.) But if (presumably through ignorance) he performs bad acts, then his power and his desire do conflict, so that the term 'voluntary' becomes no longer adequate to describe the situation. Thus it is not used in the same sense in both cases, and may be said to be an equivocal term.

The means by which Socrates has produced the final paradox of the dialogue are thus as follows: (1) the oppositions between good and bad, voluntary and involuntary, etc., (2) the equivocal use of 'good', (3) the equivocal use of 'voluntary'. Of these, (1) is not strictly a fallacious procedure, although it is an essential factor in the argument, since refutation is based on contradiction. (Hippias has wished to maintain that the bad are those who err voluntarily and the good are those who err involun-

[35] Cf. A. E. Taylor, *Plato: the Man and his Work*, p. 37:

'On reflection we see that the key to Plato's meaning is really supplied by one clause in the proposition which emerges as the conclusion of the matter: "the man who does wrong on purpose, *if there is such a person*, is the good man." The insinuation plainly is that there really is no such person as "the man who does wrong on purpose," and that the paradox does not arise simply because there is no such person.' (his italics). Cf. also Paul Shorey, *What Plato Said*, Chicago, 1958, p. 89, to the same effect. Shorey notes further, p. 471, that Plato employs a similar device at *Euthyphro* 7D ('If the gods quarrel') and at *Gorgias* 480E ('If one ought to harm anyone').

tarily; he is finally forced into a position in which he has assented to statements which imply the exact opposite.) (2) and (3) are technically fallacious, however, so that there is no doubt that the paradox has been produced by sophistical means.

What has occurred in the dialogue is this: Socrates has refuted Hippias by unfair means. Socrates knows that the refutation has been unfair and understands in what its unfairness consists. Furthermore, the paradox which completes the refutation is contrary to his own opinion, and he understands and admits this. We are now in a position to ask what Plato intends to accomplish by means of this remarkably indirect maneuver.

In order to begin to answer this question it should first be noted that the two terms which are used equivocally are both terms which are central to Socratic ethics. It may even be said, further, that the distinctive character of his ethics is expressed by means of these particular equivocations.

The two meanings of the word 'good' are, as we have seen, 'good at', or 'skilled', and 'morally good'. The good runner is the skillful runner, whereas the good man is the virtuous man. But, if virtue is to be knowledge (and there is certainly reason to believe that this is the Socratic view), there is a sense in which the two meanings will always be combined. That is, there will be a moral quality associated with good running, and a degree of technical skill associated with the performance of good deeds. Then again, the term 'voluntary' was equivocal in that it was used to mean both 'that which is in a man's power and which he also desires', and 'that which is in a man's power but which he does not desire'. (In the latter case he is deceived, through ignorance of his real interests.) This equivocation is also tied closely to one of the major Socratic doctrines: the view that no man errs voluntarily. Behind this doctrine are two assumptions: (1) that a man can do only what is in his power, and (2) of the things which are in his power he will in fact do only those which he desires, i.e., which seem good to him. The only possible explanation of error, then, is that it is done through a desire for what is not in fact good but which merely seems good. In other words, error is unknowing or involuntary. Therefore, whenever Socrates says, in his various illustrations, such things as that the good wrestler is thrown voluntarily, or that good ears are voluntarily deaf, he is, in terms of his own doctrine, uttering just as much of a paradox as he is when he states in the conclusion that it is the good man who errs voluntarily. To say that any error is voluntarily chosen shows immediately that the word 'voluntary' lacks its complete meaning in the given context; the power is present but the desire cannot be, except through the fact of an accompanying ignorance. The equivocation therefore serves to show that the whole question of error is tied to the question of knowledge, which in turn, of course, is tied to the question of virtue.

The chief point of the *Hippias Minor* therefore seems to me to be to call attention to the distinctive doctrines in the Socratic ethics. Plato has, as usual, adopted a highly individual method of effecting his purpose. He has employed arguments which, if we follow them out carefully, are clearly fallacious, yet he has used these very arguments to force us to consider the possibility that the equivocal character of certain terms may cast some light on ethical questions of the most fundamental sort. There are of course other things accomplished in the dialogue: we see, for instance, that Hippias is hopeless as a dialectician, that the use of exhaustive alternatives is essential to the refutation of a statement, and that equivocation is a fallacy which Plato handles with ease. But Plato's real accomplishment in the dialogue is, by means of a paradox which is a shock to our common sense, to compel us to consider what the terms 'good' and 'voluntary' really mean. This is a use of fallacy which serves to construct, not, as in the *Euthydemus*, to destroy.[36] If the constructive nature of the arguments is not as obvious as in the *Theaetetus*, this is because Hippias, as a dramatic character, is not capable of making the distinctions which would bring this out. Moreover, Socrates presumably has little interest in

[36] Aristotle writes that the solutions of fallacious arguments are

> useful for philosophy for two reasons. In the first place, as they generally turn on language, they put us in a better position to appreciate the various meanings which a term can have and what similarities and differences attach to things and their names. Secondly they are useful for the questions which arise in one's own mind; for he who is easily led astray by another person into false reasoning and does not notice his error, might also often fall into this error in his own mind.

(*On Sophistical Refutations*, 175a5ff., trans. Forster.) I should guess that Plato would have agreed with both points.

the education of Hippias, so that Plato does not represent him as making the distinctions either. But the necessary hints are there, and it is the reader's responsibility to heed them.[37]

[37] The authenticity of the *Hippias Minor* is attested to by Aristotle at *Metaphysics* Δ 1025a6ff. in the course of his discussion of the meanings of 'false' (ψεῦδος):

Hence the proof in the *Hippias* that the same man is false and true is misleading; for its assumes (*a*) that the false man is he who is *able* to deceive, i.e. the man who knows and is intelligent; (*b*) that the man who is willingly bad is better. This false assumption is due to the induction; (διὰ τῆς ἐπαγωγῆς) for when he says that the man who limps willingly is better than he who does so unwillingly, he means by limping *pretending* to limp. For if he is willingly lame, he is presumably worse in this case just as he is in the case of moral character (trans. Tredennick).

Aristotle's comment is worth considering on its own account. The passage taken as a whole illustrates the fact that Aristotle conceives the dialogue as consisting, as I have said, of a single argument, since he cites as contributing to the first refutation (that the same man is false and true) not only (*a*), which is drawn from the first discussion (especially 366A), but also (*b*), which is drawn from the second (especially 374B). In other words, he takes the conclusion that the same man is false and true as substantially equivalent to the conclusion that the good man errs voluntarily. (This could be rephrased as 'the same man is both good and bad'.) Further in (*a*), Aristotle notes the connection between power and knowledge and relates this to the first refutation. (We saw that the notion of power, when first introduced, was understood by Hippias to mean power for evil; the introduction of shrewdness 365E prepared the way for a shift to a power connected with good, i.e., with knowledge. It was by means of this shift that the false man was identified with the true.) In (*b*), Aristotle has an interesting comment to make on the argument which Socrates has used to reach the conclusion that 'the man who is willingly bad is better'. This false assumption, he says, is due to the induction; by this I take him to mean that the example about voluntary bodily error should not have been used to lead to a conclusion involving voluntary moral error since the two types of activity are not strictly comparable. (That is, in the example about limping a pretense is made, which seems not to be the case in the conclusion.) If the man were in fact willingly lame (I think Aristotle means through having deliberately inflicted an injury upon himself) the induction would hold in the sense that bodily and moral error could be compared, but the conclusion would be that the man committing moral error would be worse, not better. Aristotle, it must be remembered, rejects the Socratic paradox that no man errs voluntarily (see e.g., *Nichomachean Ethics* 1136a1 ff.); thus we have a situation in which he and Plato, although holding opposite views on the question of voluntary error, come both to reject the conclusion (*b*) (that the man who is willingly bad is better): Aristotle because he objects to 'better', Plato because he objects to 'willingly bad'.

Conclusion

IN the preceding chapters I have been chiefly concerned with some of the ways in which Plato has made use of fallacy in connection with the exposition of his own philosophical views. But I have also asserted (1) that Plato possesses a complete mastery of certain kinds of fallacy (especially the fallacies of equivocation and *secundum quid*), and (2) that on occasion he deliberately employs such fallacies as an indirect means of supporting his own position.

Of these two points, the first is more immediately palatable than the second. It is one thing to credit Plato with an understanding of certain points of logic, but quite another to assert that this understanding is put to what might be regarded as questionable use. If fallacies are, after all, bad arguments, it might be asked whether it is fair play on the part of Plato to employ them, even for the most laudable ends.

There is a real dilemma here. In the first place, there is certainly no doubt that fallacies do occur in the dialogues. If so (to argue for a moment in Eleatic style), Plato either knew this or he did not. If he knew it, he lays himself (and Socrates) open to the charge of immoral practise in argument; if he did not, he is apparently ignorant of some fairly elementary parts of logic.[38]

[38] Sometimes, of course, it is argued that Plato could not have been expected to know much logic at this stage of the world's history. See, for instance, E. R. Dodds in his commentary on *Gorgias* 474C4–476A2 (Oxford, 1959, p. 249) who takes this line, but not without a certain uneasiness:

But while Polus' view is muddled and ultimately untenable, Socrates' formal 'refutation' of it seems to turn merely on the ambiguity of the word ὠφέλιμον. When Polus said that doing wrong was less admirable, he clearly meant that it was

In this book I have argued that the evidence of the *Euthydemus* is too strongly on the side of Plato's conscious use of fallacy for his understanding of the fallacies in that dialogue to be denied. I have then proceeded to argue that when similar fallacies occur in subsequent dialogues, they continue to be understood and their use is still deliberate. On the other hand, I have also argued (particularly at the close of the *Theaetetus* chapter, pp. 43ff.) that arguments which may, from their resemblance in form to arguments in the *Euthydemus*, be called 'eristic', need by no means be employed in an eristic spirit. In this way I have tried to uphold Plato's intelligence without representing him as a logical twister, but I am actually less concerned to defend his character than to describe what he appears to be doing in certain dialogues. Whether what he does is reprehensible can only be decided by the individual reader, on inspection of the passages involved. The whole issue would in fact be trivial if it were not that I suspect strongly that it is precisely this reluctance to pass an adverse moral judgment upon Plato (or perhaps more especially upon Socrates) which has prevented scholars from considering seriously the possibility that Plato's use of fallacy really is deliberate. The result has been that Plato's competence as a logician has failed to be evaluated correctly—a matter which is far from trivial.[39]

less ὠφέλιμον *for the community* and from this it does not immediately follow that it is less ὠφέλιμον *for the agent*, i.e. κάκιον in Polus' sense of that term. The underlying thought is, no doubt, that since regard for justice is recognized even by Polus as καλόν, as evoking immediate admiration, it must be a necessary constituent of the 'good' or happy life. But Plato has obscured this point for the sake of giving his argument the appearance of a formal proof. We must remember that when the *Gorgias* was written the study of logic was still in its earliest infancy (as Aristotle's *Sophistici Elenchi* sufficiently shows). Nevertheless, it is not easy to believe with T. Gomperz and others that Plato was wholly unconscious of the equivocation . . . .

[39] Of course it is doubtful whether the Greeks themselves would have been much troubled by the moral issue. E.g., Aristotle in *On Sophistical Refutations* (a treatise in which, incidentally, he gives some excellent advice on how to be a sophist of the eristic type, 172b10–173a31 and 174a17–174b40) writes as follows:

In the first place, then, just as we say that we ought sometimes deliberately to argue plausibly rather than truthfully, so too we ought sometimes to solve questions plausibly rather than according to truth. For, generally speaking, when we have to fight against contentious arguers, we ought to regard them not as trying to refute us but as merely appearing to do so; for we deny that they are arguing a case, so that they must be corrected so as not to appear to be doing so. For if

As I have said in the introduction, the question of Plato's consciousness of the fallacious character of a given argument needs to be settled before proceeding to a discussion of the way in which that fallacy is used. But it may be of interest, at the conclusion of this study, to examine brief passages in additional dialogues in which it seems highly probable, if not completely settled, that Plato knew quite well what he was doing.

I. *Laches* 190Bff.

The dialogue has begun with the question whether the art of fighting in armor should be learned by young men, but has now come round to inquire about the nature of virtue 190B. It will be easier, Socrates says, to investigate a part of virtue rather than the whole—why not consider the particular virtue which the art of fighting in armor is supposed to produce, i.e., courage? Laches is asked to tell what courage is, but, like many of those who converse with Socrates, he does not at first understand the nature of a general definition. After one false start, however, he suggests that courage is 'a sort of endurance of the soul' (*καρτερία τις τῆς ψυχῆς* 192B). Socrates argues against the proposed definition in the following way: courage presumably is a noble thing. With respect to endurance, however, wise endurance is good and noble, but foolish endurance evil and hurtful. It seems then, as though only wise endurance should be courage. Yet there appear to be cases in which we regard a man who endures foolishly (e.g., a soldier in the face of overwhelming odds) as braver than the man who endures with sense. To call

refutation is unequivocal contradiction based on certain premisses, there can be no necessity to make distinctions against ambiguity and equivocation; for they do not make up the proof. But the only other reason for making further distinctions is because the conclusion looks like a refutation. One must, therefore, beware not of being refuted but of appearing to be so . . . (175a32ff., trans. Forster).

Furthermore it seems clear that Aristotle regarded the respondent in a dialectical argument as entirely responsible for his own defense: 'since the right to draw distinctions is conceded in arguments, it is obvious that to grant the question simply, without making distinctions, is a mistake' 175b30ff. The opponents of Socrates, when they are refuted, quite regularly are so because of this very failure to draw the right distinctions; we need not fail to draw them, however, since Plato has on most occasions provided the means for us to do so.

this man 'courageous' would contradict the original statement that courage is noble but foolish endurance is ignoble.

Laches admits the contradiction but can see no way of escape. There is in fact no escape so long as he continues to call 'courageous' actions which should more properly be called 'rash'. After Nicias enters the discussion, however, this distinction is ultimately made.

Nicias offers (195A, 196C) a definition of courage as 'knowledge of the grounds of hope and fear' (ἡ τῶν δεινῶν καὶ θαρραλέων ἐπιστήμη). Among other arguments against this definition, Socrates puts forward one which is basically the same as part of his objection to the definition offered by Laches:

> ... I conceive it is necessary for him who states this theory to refuse courage to any wild beast, or else to admit that a beast like a lion or a leopard or even a boar is so wise as to know what only a few men know because it is so hard to perceive. Why, he who subscribes to your account of courage must needs agree that a lion, a stag, a bull, and a monkey have all an equal share of courage in their nature.
>
> *Lach.* Heavens, Socrates, how admirably you argue! Now answer us sincerely, Nicias, and say whether those animals, which we all admit to be courageous, are wiser than we are; or whether you dare, in contradiction of everyone else, describe them as not even courageous. (196E.)

Nicias, in response, makes the answer which Laches should have made in the case of the foolhardy soldier, etc., in the earlier passage;

> No, Laches, I do not describe animals, or anything else that from thoughtlessness has no fear of the dreadful, as courageous, but rather as fearless and foolish. Or do you suppose I describe all children as courageous, that have no fear because they are thoughtless? I rather hold that the fearless and the courageous are not the same thing. . . . So you see, the acts which you and most people call courageous, I call rash, and it is the prudent acts which I speak of that are courageous. (197Aff.)

Socrates has succeeded in pushing the popular (equivocal) use of 'courage' so far that he has at last forced his opponent to separate off one meaning of the term, i.e. rashness. (In the end, courage becomes a divine type of knowledge and is thus not simply a part but the whole of virtue 199E.) The way in which he has achieved the separation is by ignoring it himself and by proceeding as if the popular use of courage were the correct one. Laches can see that something is wrong, he is not sure what; Nicias sees the trouble and states it clearly. Thus deliberate ambiguity on the part of Socrates results in the clarification of this same ambiguity. The dialogue, like others of the same period, finishes with an admission of defeat 199E, but it is certainly not the case that nothing has been accomplished, since we have at least been made to see that if courage is a virtue it cannot be foolish. This may result in the absorption of courage into virtue as a whole (the problem of the *Protagoras*) but progress has been made in the direction of a distinction between popular courage and the courage proper to the philosopher. Plato has made use of the fallacy of equivocation in a constructive way.[40]

II *Meno* 80Dff.

Even after some discussion of the nature of virtue, Socrates still does not know what virtue is. He asks Meno to join him in the inquiry, but Meno has an objection to the project:

Why, on what lines will you look, Socrates, for a thing of whose nature you know nothing at all? Pray, what sort of thing, amongst those that you know not, will you treat us to as the object of your search? Or even supposing, at the best, that you hit upon it, how will you know it is the thing you did not know?

Socrates is quite familiar with this sort of thing:

I understand the point you would make, Meno. Do you see what a captious argument you are introducing—that, forsooth, a man cannot inquire either about what he knows or about what he does not know? For he cannot inquire about what he knows, because he

[40] Plato fairly frequently employs the method we have just seen Socrates using here in the *Laches*, that of taking a term in such a wide sense initially that the opponent will finally insist on having the extension limited. We may compare *Gorgias* 489B, C where Socrates forces Callicles to say that by 'superior' (κρείττων) he does not mean 'physically stronger', and *Republic* I 338C, D where a similar treatment is administered to Thrasymachus. (In these cases, however, the limitation of meaning results in clarification of the opponent's wrong view, not that of Socrates.) In his note on the latter passage Shorey writes, 'To the misunderstanding of such dramatic passages is due the impression of hasty readers that Plato is a sophist' (Loeb edition, *ad loc.*).

knows it, and in that case is in no need of inquiry; nor again can he inquire about what he does not know, since he does not know about what he is to inquire.

Men. Now does it seem to you to be a good argument, Socrates?
Soc. It does not.
Men. Can you explain how not?
Soc. I can. . . . (80Dff.)

Socrates then goes on to outline the doctrine of recollection (ἀνάμνησις), throwing special emphasis upon the fact that this doctrine is one which makes learning possible:

For as all nature is akin, and the soul has learned all things, there is no reason why we should not, by remembering but one single thing —an act which men call learning—discover everything else, if we have courage and faint not in the search; since, it would seem, research and learning are wholly recollection. So we must not hearken to that captious argument: it would make us idle, and is pleasing only to the indolent ear, whereas the other makes us energetic and inquiring. Putting my trust in its truth, I am ready to inquire with you into the nature of virtue. (81Cff.)

The questioning of the slave boy then follows by way of illustration. The investigation of the nature of virtue, however, does not proceed until Socrates has again spoken out against the evil effects of Meno's argument in the strongest terms:

. . . Most of the points I have made in support of my argument are not such as I can confidently assert; but that the belief in the duty of inquiring after what we do not know will make us better and braver and less helpless than the notion that there is not even a possibility of discovering what we do not know, nor any duty of inquiring after it—this is a point for which I am determined to do battle, so far as I am able, both in word and deed. (86Bff.)

Like arguments in both the *Euthydemus* and the *Theaetetus*, this argument of Meno's at 80D is based upon an Eleatic distinction between knowing and not knowing which is ultimately destructive of the possibility of acquiring any new knowledge. The doctrine of recollection is introduced as a direct answer to an eristic trick, a trick which would stultify the desire to learn through the assertion that learning is impossible. The doctrine of recollection is, of course, as may be seen more clearly in the *Phaedo*, closely connected with the theory of Forms, so that the fact that Plato chooses to meet the eristic argument in the way he does is further evidence of the constant clash between the Eleatic metaphysics and the Platonic. Fallacy has been used by Plato, as it so often is, to give emphasis to some important positive doctrine of his own.[41]

The way in which the fallacy is introduced, however, differs from the method of the *Laches* in two important ways: (1) It occurs not in the mouth of Socrates but of one of his interlocutors. (In this respect the passage may be compared with *Cratylus* 429Bff. where it is Cratylus, not Socrates, who puts forward the arguments against false speaking.) (2) It is immediately recognized as a fallacy and tagged as such. The result is that the difficulties about consciousness and culpability really do not arise.

The fact seems to be that the acceptance of the thesis that Plato's use of fallacy is deliberate presents difficulties chiefly when fallacious argument occurs in the mouth of Socrates (or of some other speaker whom we tend also to regard as spokesman for Plato, e.g. the Eleatic Stranger). This situation is a natural by-product of the dialogue form. One's instinct when reading the account of an argument is to take sides, and to assume, further, that the author has taken sides also. In the case of a Socratic dialogue, it seems obvious that the side taken by the author will be that of Socrates. If, then, we find Socrates (in spite of his supposedly good character) tricking an opponent by fallacious means, we may be inclined to shift our sympathies to the other side, to the side of the person being tricked. (Or, as I have already suggested, if we prefer to think that Socrates would not consciously behave in such a way, we may decide instead that he, and probably Plato also, were ignorant of the logical issues involved.) But all this taking of sides, although a tribute to Plato's dramatic skill, is really beside the point. The way to read a Socratic dialogue is not only to ask one's self, what

[41] The philosophical importance of this passage in the *Meno* is well brought out by Bernard Phillips in an article 'The Significance of Meno's Paradox' (*Classical Weekly*, vol. 42, no. 6, December 20, 1948, pp. 87–91). He writes, for instance,

> The paradox is evidently of sophistic origin, and though in the hands of lesser sophists and laymen it served only as a piece of verbal trickery, the argument itself expresses a philosophical point of view which is opposed fundamentally to the whole conception of the nature of knowledge and the function of reason which Plato is advancing.

would I have said if I had been in the place of Hippias or Cratylus or Protagoras, but also (and this is more important), what has Plato actually meant to say to me, the reader, by means of the entire conversation? If we read the dialogues in this way, there is much less likelihood of confusing ourselves by extraneous moral judgments, for instance: 'Plato never gives Hippias anything intelligent to say, so naturally Socrates wins the argument; all this is very unfair.' Of course if Plato had written prose treatises, none of these problems would arise, but he did not write prose treatises for the very good reason that he conceived the exercise of philosophy to be the exercise of dialectic—hence the only way to give even a semblance of philosophical activity in written form was to write dialogues. (That he did write dialogues is the Platonic scholar's joy and also his despair.)

The whole question of Plato's use of fallacy is one which is unusually closely connected with dialectic and hence with the dialogue form, since (1) it is in the process of conversation that the various instances of fallacious argument almost invariably occur, and (2) fallacy is often part of the ἔλεγχος, of the dialectical shock-treatment administered by Socrates as the torpedo-fish. It is for this reason that I have tried to study Plato's fallacies as they appear in complete dialogues or at least in fairly complete conversations, and also why I have tried to take into account the aims and capacities of the various speakers Plato has presented to us.

The chief generalization which I would wish to make in conclusion is, then, the one which has already been suggested in the introduction: that Plato, when he employs fallacious argument, does so in order to draw our attention to some fundamental philosophical problem. We have now seen what some of these problems are: the meaning of terms like 'good', 'voluntary', 'learn', the difficulties in the Eleatic either-or, the contrast between Eleaticism and the theory of Forms, the problems of participation, of imitation, of learning and memory, of otherness, of the possibility of knowledge. Such a list shows that the study of fallacy in Plato cannot be separated from a study of the questions most central to his thought; it is therefore not a topic to be neglected.

Appendix

FR. BOCHEŃSKI ON PLATO'S LOGIC

IN Fr. Bocheński's brief survey, *Ancient Formal Logic* (Amsterdam, 1951), Plato receives credit for the following contributions: (1) some remarks on syntax in the *Sophist* (probably not original with Plato) (p. 14); (2) an elementary semantics in the *Cratylus* (probably also not original) (p. 15); (3) an analogon of $\ulcorner Ax \supset {\sim}Ax \,.\supset.\, {\sim}Ax \urcorner$ at *Theaetetus* 171A (p. 16);[42] (4) the first conception and clear statement of 'the ideal of valid laws of reasoning' at *Timaeus* 47B (p. 18); (5) a shift in attention 'from the apagogic proofs to positive demonstrations of statements attributing a propriety to a subject' which paved the way for the logic of terms (p. 18); (6) the method of division, which became the origin of the syllogism (p. 18); (7) the rudiments of almost everything in Aristotle's logic with the exception of the analytical syllogism and some related doctrines (p. 18).

On the other hand, Bocheński has also this to say of Plato's logic:

> His dialectics appears to us as being a confusion of different sciences and different methods. It includes on the one hand the art of disputation, metaphysics and logic; on the other hand Plato does not distinguish between formal logic, methodology (of a kind rather akin to that of empirical sciences of today) and the intuitive approach to (mostly) axiological problems. *The reading of his dialogues is almost intolerable to a logician, so many elementary blunders are contained in them.* It will be enough to mention his struggling with the false principle $\ulcorner SaP \supset \bar{S}a\bar{P} \urcorner$ [e.g. *Gorgias* 507A; Alcibiades I 126C][43] or the difficulty he has in grasping that who does not admit $\ulcorner SaP \urcorner$ must not necessarily admit $\ulcorner SeP \urcorner$. [E.g. *Gorgias* 466A[44];

[42] See also *Formale Logik*, Freiburg/München, 1956, pp. 38–9.

[43] I have looked at *Alcibiades* I 126C a number of times, but I must confess I do not see its relevance in this context.

[44] At *Gorgias* 466A Polus tries to assert that rhetoric is flattery rather than, what Socrates has been at pains to assert, that it is a branch (*μόριον*) of

Meno 73E[45]; *Protagoras* 350ff.[46]] (p. 17). . . . *Correct logic we find none in his work*; he was, however, a thinker who during his whole life was searching for logic and paved the way for its founder (p. 18). (My italics.)

Bocheński again discusses Plato's logic in his much larger *Formale Logik* (Freiburg/München, 1956, pp. 39–46). He again credits Plato with the first clear concept of logic (using the same *Timaeus* passage as before),[47] and with the method of division (which he here illustrates with parts of *Sophist* 218D–221C). There are no remarks about logical blunders of the sort italicized above, but so much emphasis is laid upon the immense difficulty which Plato experienced in arriving at the most simple logical laws that it is hard to feel that Bocheński wishes to give Plato very much credit in cases of his ultimate success:

Wie schwer ihn die Lösung der uns als elementar anmutenden Fragen der Logik ankam, mögen die folgenden Stücke seiner Dialektik nahebringen: er ringt in ihnen mit Mühe um ganz einfache Gesetze (p. 40).[48]

The illustrative passages are *Gorgias* 507A, *Euthyphro* 12A, and *Protagoras* 350Cff. (The *Euthyphro* passage is new, whereas *Gorgias* 466A, *Meno* 73E, and *Alcibiades* I 126C have been dropped.) Bocheński then writes:

Im ersten dieser Texte handelt es sich um die (falsche) These: Falls, wenn *A* dem *x* zukommt, auch *B* dem *x* zukommt, dann: falls *A* dem *x* nicht

---

flattery. There seems to be no question here of anyone wishing to infer ⌜SeP⌝ from the denial of ⌜SaP⌝. Rather Plato is simply pointing out that flattery is a wider concept than rhetoric.

45 *Meno* 73E appears to be similar to *Gorgias* 466A. Here, when Meno says that justice is virtue, Socrates asks, 'Virtue or *a* virtue?' They then agree that it must be *a* virtue since there are other virtues beside justice; virtue is wider than justice, in other words. (Plato does not say so but I would guess he means also to imply that the A proposition 'All justice is virtue' is therefore not convertible *simpliciter*, and so for 'All rhetoric is flattery' in the *Gorgias*. See below on *Euthyphro* 12Aff.)

46 At *Protagoras* 350ff. the question is whether ⌜SaP⌝ implies ⌜PaS⌝. (Plato is clear that it does not.) But see below pp. 94–96.

47 I should certainly like to see the clear concept of logic which Bocheński sees in this passage, but I honestly do not. In commenting on it he writes 'Ohne den Begriff des allgemeingültigen Gesetzes war offenbar keine formale Logik möglich' (p. 40), and with this I should certainly agree. But λογισμοί 47C are surely calculations or reasonings, not laws.

48 'The following extracts from his dialectic, in which he makes a laboured approach to quite simple laws, show how difficult he found it to solve logical questions that seem elementary to us.' (Translation by Ivo Thomas of *Formale Logik* with the title *A History of Formal Logic*, Indiana, 1961, p. 34.)

zukommt, dann kommt *B* dem *x* nicht zu. Der zweite zeigt die Schwierigkeiten um die Konvertibilität des Allsatzes: ob nämlich aus ‚alle *A* sind *B*' auch ‚alle *B* sind *A*' folgt. Wie schwer diese Frage Platon fiel, zeigt in noch klarerer Weise der dritte Text; er ist zudem deshalb höchst interessant, weil Platon darin, um die Ungültigkeit der genannten Konversionsregel aufzuweisen, zu komplizierten ausser-logischen Erörterungen (wie etwa zu derjenigen über die körperliche Stärke) seine Zuflucht nimmt (p. 42).[49]

Since the three texts referred to here seem to be the ones by which Bocheński sets special store, let us examine each of them in turn.

(1) *Gorgias* 507A

ΣΩ. Λέγω δὴ ὅτι, εἰ ἡ σώφρων [sc. *ψυχὴ*] *ἀγαθή ἐστιν, ἡ τοὐναντίον τῇ σώφρονι πεπονθυῖα κακή ἐστιν· ἦν δὲ αὕτη ἡ ἄφρων τε καὶ ἀκόλαστος. Πάνυ γε. Καὶ μὴν ὅ γε σώφρων τὰ προσήκοντα πράττοι ἂν καὶ περὶ θεοὺς καὶ περὶ ἀνθρώπους· οὐ γὰρ ἂν σωφρονοῖ τὰ μὴ προσήκοντα πράττων. Ἀνάγκη ταῦτ' εἶναι οὕτως.*

Soc. I say, then, that if the temperate soul is good, one that is in the opposite state to this sensible one is bad; and that was the senseless and dissolute one. Certainly. And further, the sensible man will do what is fitting as regards both gods and men; for he could not be sensible if he did what was unfitting. That must needs be so.

The first sentence of this passage certainly appears to state that if the temperate soul is good, the soul in the opposite condition (i.e., the intemperate soul) is bad. Therefore, as Bocheński writes, this 'involves' the false principle that $\ulcorner Ax \supset Bx \,.\supset.\, \sim Ax \supset \sim Bx \urcorner$. Bocheński does not explicitly state that the passage should be cited as an example of incorrect logic on Plato's part, but the reader certainly receives the impression that the quotation was not given with the idea of paying the philosopher a compliment. It would seem to me that although Bocheński is right to point out that the passage has some relevance to the false principle he has indicated, there is need to study it at considerably greater length before it could become useful to cite it as evidence either in Plato's favor or the reverse. (Just how much praise or blame Bocheński attaches to the

[49] 'In the first of these texts is involved the (false) thesis: Suppose, if *A* belongs to *x*, *B* also belongs to *x*, then: if *A* does not belong to *x*, then *B* does not belong to *x*. The second shows the difficulties found concerning the convertibility of universal affirmative sentences: viz. whether "all B is A" follows from "all A is B". The third text shows still more clearly how hard Plato felt these questions to be; it further has the great interest that, to show the invalidity of the fore-going rule of conversion, he betakes himself to complicated extra-logical discussions—about bodily strength, for instance' (trans. Thomas, p. 35).

passage is not completely clear.) Further study of the passage might be conducted along the following lines:

(*a*) Is it possible that ἡ σώφρων[ψυχὴ]ἀγαθή should be translated not, 'the temperate soul is good', but 'the temperate soul is the good soul'? If Plato means that the temperate soul is a member of the class of good things, his inference is invalid. If, however, he means that '*x* is a temperate soul' and '*x* is a good soul' are equivalent, then his inference is correct. (He would have asserted $\ulcorner Bx \supset Ax \urcorner$ as well as $\ulcorner Ax \supset Bx \urcorner$; with this additional assertion, the inference $\ulcorner \sim Ax \supset \sim Bx \urcorner$ becomes valid.)

(*b*) If Plato had added 'the good soul is temperate' his inference would have been valid since $\ulcorner Ax \supset Bx . Bx \supset Ax . \supset . \sim Ax \supset \sim Bx \urcorner$. It is possible that Plato regarded this assertion as obvious and thus did not trouble to make it? (By this I mean to suggest that, in Plato, logic may well be affected by literary form.)

(*c*) Assuming that the inference is invalid, it is possible that Plato is conscious of the fallacy and is employing it deliberately? 507A is part of a very long argument, and I shall not attempt to study that whole argument here. But in view of the general thesis of this book, this seems to me a possibility worth considering.

(*d*) Again assuming that the inference is invalid, what are we to make of the very next sentence in the same paragraph? For this sentence, as far as I can see, is an instance of the correct principle $\ulcorner Ax \supset Bx . \supset . \sim Bx \supset \sim Ax \urcorner$ (if the sensible man does what is fitting, then a man who does what is not fitting is not sensible). We can surely not be asked to believe that Plato gets an inference wrong in one sentence and right in the next—unless of course Bocheński means to cite the passage purely as an example of Plato's struggles with simple logical laws, or unless we are to assume, which is possible, but seems to me unlikely, that Plato may have thought both $\ulcorner Ax \supset Bx . \supset . \sim Ax \supset \sim Bx \urcorner$ *and* $\ulcorner Ax \supset Bx . \supset . \sim Bx \supset \sim Ax \urcorner$.

(*e*) Assuming once more that the inference is invalid, what are we to make of passages in other dialogues in which Plato makes similar inferences correctly? For instance:

(i) *Hippias Minor* 366B

ΣΩ. Ὡς ἐν κεφαλαίῳ ἄρα εἰρῆσθαι, οἱ ψευδεῖς εἰσιν οἱ σοφοί τε καὶ δυνατοὶ ψεύδεσθαι.
ΙΠ. Ναί.
ΣΩ. Ἀδύνατος ἄρα ψεύδεσθαι ἀνὴρ καὶ ἀμαθὴς οὐκ ἂν εἴη ψευδής.
ΙΠ. Ἔχει οὕτως.

Soc. In short, then, the false are those who are wise and powerful in uttering falsehoods.

Hipp. Yes.

Soc. A man, then, who has not the power to utter falsehoods and is ignorant would not be false.

Hipp. That is true.

Like the second sentence of the *Gorgias* passage, this is an instance of the correct principle $\ulcorner Ax \supset Bx . \supset . \sim Bx \supset \sim Ax \urcorner$.

(ii) *Euthyphro* 7A

ΕΥ. *῎Εστι τοίνυν τὸ μὲν τοῖς θεοῖς προσφιλὲς ὅσιον, τὸ δὲ μὴ προσφιλὲς ἀνόσιον.*

Eu. Well then, what is dear to the gods is holy, and what is not dear to them is unholy.

This passage appears to contain the same mistake as the first sentence of the *Gorgias* passage. However, if we follow the argument along to the point of its refutation at 10Eff., we shall find that Socrates has been regarding Euthyphro's proposed definition as a biconditional not as a case of class inclusion. In this case, the inference at 7A would have been correct.

(2) *Euthyphro* 12A

ΣΩ. . . . *ἰδὲ γάρ, εἰ οὐκ ἀναγκαῖόν σοι δοκεῖ δίκαιον εἶναι πᾶν τὸ ὅσιον.*

ΕΥ. *῎Εμοιγε.*

ΣΩ. *᾿Αρ᾽ οὖν καὶ πᾶν τὸ δίκαιον ὅσιον ἢ τὸ μὲν ὅσιον πᾶν δίκαιον, τὸ δὲ δίκαιον οὐ πᾶν ὅσιον, ἀλλὰ τὸ μὲν αὐτοῦ ὅσιον, τὸ δέ τι καὶ ἄλλο;*

ΕΥ. *Οὐχ ἕπομαι, ὦ Σώκρατες, τοῖς λεγομένοις.*

Soc. . . . Just see whether you do not think that everything that is holy is right.

Eu. I do.

Soc. But is everything that is right also holy? Or is all which is holy right, and not all which is right holy, but part of it holy and part something else?

Eu. I can't follow you, Socrates.

Quite far from seeing in this passage evidence that Plato had difficulty with the correct conversion of universal propositions, I see here evidence that he had no such difficulty but understood perfectly that an A proposition converts not *simpliciter*, but *per accidens*. Plato's complete mastery of this fact becomes even more evident when we see him immediately going on to discuss another A proposition, 'all fear is reverence'. This Socrates rejects since he can think of cases in which men fear things which they do not reverence.

(An O proposition contradicts an A proposition, in other words.) However, he thinks that the converse, 'all reverence is fear', will hold, and this statement is perfectly compatible with 'some fear is not reverence'. The reason why these two statements may be asserted together whereas it would not be correct to make a joint assertion consisting of 'all fear is reverence' and 'some fear is not reverence' could hardly be explained more clearly than it is by Plato at 12C:

ΣΩ. *Οὐκ ἄρ' ὀρθῶς ἔχει λέγειν· ἵνα γὰρ δέος, ἔνθα καὶ αἰδώς· ἀλλ' ἵνα μὲν αἰδώς, ἔνθα καὶ δέος, οὐ μέντοι ἵνα γε δέος, πανταχοῦ αἰδώς. ἐπὶ πλέον γάρ, οἶμαι, δέος αἰδοῦς· μόριον γὰρ αἰδὼς δέους, ὥσπερ ἀριθμοῦ περιττόν, ὥστε οὐχ ἵνα περ ἀριθμός, ἔνθα καὶ περιττόν, ἵνα δε περιττόν, ἔνθα καὶ ἀριθμός. ἔπει γάρ που νῦν γε;*
ΕΥ. *Πάνυ γε.*

Soc. Then it is not correct to say 'where fear is, there also is reverence'. On the contrary, where reverence is, there also is fear; but reverence is not everywhere where fear is, as I think, fear is more comprehensive than reverence; for reverence is a part of fear, just as the odd is a part of number, so that it is not true that where number is there also is the odd, but that where the odd is there also is number. Perhaps you follow me now?
Eu. Perfectly.

In other words, Socrates is here saying that where one concept is wider (*πλέον*) than another, it is correct to predicate this wider concept of the narrower one but incorrect to predicate the narrower concept of the wider. 'Fear', in this instance, is a wider concept than reverence, since although everything which is reverenced is also feared, there are some things which are feared but not reverenced. When Euthyphro has understood this point, it is then possible for Socrates to return to his original point about the holy and the right:

ΣΩ. *Τὸ τοιοῦτον τοίνυν καὶ ἐκεῖ λέγων ἠρώτων, ἆρα ἵνα δίκαιον, ἔνθα καὶ ὅσιον, ἢ ἵνα μὲν ὅσιον, ἔνθα καὶ δίκαιον, ἵνα δὲ δίκαιον, οὐ πανταχοῦ ὅσιον· μόριον γὰρ τοῦ δικαίου τὸ ὅσιον. οὕτω φῶμεν ἢ ἄλλως σοι δοκεῖ;*
ΕΥ. *Οὔκ, ἀλλ' οὕτω. φαίνει γάρ μοι ὀρθῶς λέγειν.*

Soc. It was something of this sort that I meant before, when I asked whether where the right is, there also is holiness, or where holiness is, there also is the right; for holiness is not everywhere where the right is, for holiness is a part of the right. Do you agree to this, or do you dissent?
Euth. No, I agree; for I think the statement is correct. (12Cff.)

Here it is plain that Plato means to establish the fact that 'right' is a wider concept than 'holy' in the same way that 'fear' is a wider concept than 'reverence'. Since he has already pointed out that the latter pair is not convertible, and has done this for the sake of illustration, it hardly seems possible that he could have thought the former pair to be convertible.

However, I am perhaps misunderstanding the inference which Bocheński wishes to have drawn from his use of *Euthyphro* 12A as an illustration of Plato's difficulties with the convertibility of universals; perhaps the reader is intended to infer that although Plato found the question difficult, his struggles met with success. Even so, the general impression given is that if Plato had to work so hard to formulate such a very elementary rule, he must have been rather particularly stupid. Thus the formulation of the rule, even if achieved after this extraordinary expenditure of effort, would seem to be regarded as of no great value.

(3) *Protagoras* 350C–351B

ΠΡΩ. . . . ἔγωγε ἐρωτηθεὶς ὑπὸ σοῦ, εἰ οἱ ἀνδρεῖοι θαρραλέοι εἰσίν, ὡμολόγησα· εἰ δὲ καὶ οἱ θαρραλέοι ἀνδρεῖοι, οὐκ ἠρωτήθην· εἰ γάρ με τότε ἤρου, εἶπον ἂν ὅτι οὐ πάντες· τοὺς δὲ ἀνδρείους ὡς οὐ θαρραλέοι εἰσί, τὸ ἐμὸν ὁμολόγημα οὐδαμοῦ ἐπέδειξας ὡς οὐκ ὀρθῶς ὡμολόγησα. ἔπειτα τοὺς ἐπισταμένους αὐτοὺς ἑαυτῶν θαρραλεωτέρους ὄντας ἀποφαίνεις καὶ μὴ ἐπισταμένων ἄλλων, καὶ ἐν τούτῳ οἴει τὴν ἀνδρείαν καὶ τὴν σοφίαν ταὐτὸν εἶναι· τούτῳ δὲ τῷ τρόπῳ μετιὼν καὶ τὴν ἰσχὺν οἰηθείης ἂν εἶναι σοφίαν. πρῶτον μὲν γὰρ εἰ οὕτω μετιὼν ἔροιό με εἰ οἱ ἰσχυροὶ δυνατοί εἰσι, φαίην ἄν: ἔπειτα, εἰ οἱ ἐπιστάμενοι παλαίειν δυνατώτεροί εἰσι τῶν μὴ ἐπισταμένων παλαίειν καὶ αὐτοὶ αὐτῶν, ἐπειδὰν μάθωσιν, ἢ πρὶν μαθεῖν, φαίην ἄν· ταῦτα δὲ ἐμοῦ ὁμολογήσαντος ἐξείη ἄν σοι, χρωμένῳ τοῖς αὐτοῖς τεκμηρίοις τούτοις, λέγειν ὡς κατὰ τὴν ἐμὴν ὁμολογίαν ἡ σοφία ἐστὶν ἰσχύς. ἐγὼ δὲ οὐδαμοῦ οὐδ' ἐνταῦθα ὁμολογῶ τοὺς δυνατοὺς ἰσχυροὺς εἶναι, τοὺς μέντοι ἰσχυροὺς δυνατούς: οὐ γὰρ ταὐτὸν εἶναι δύναμίν τε καὶ ἰσχύν, ἀλλὰ τὸ μὲν καὶ ἀπὸ ἐπιστήμης γίγνεσθαι, τὴν δύναμιν, καὶ ἀπὸ μανίας γε καὶ θυμοῦ, ἰσχὺν δὲ ἀπὸ φύσεως καὶ εὐτροφίας τῶν σωμάτων. οὕτω δὲ κἀκεῖ οὐ ταὐτὸν εἶναι θάρσος τε καὶ ἀνδρείαν· ὥστε συμβαίνει τοὺς μὲν ἀνδρείους θαρραλέους εἶναι, μὴ μέντοι τούς γε θαρραλέους ἀνδρείους πάντας· θάρσος μὲν γὰρ καὶ ἀπὸ τέχνης γίγνεται ἀνθρώποις καὶ ἀπὸ θυμοῦ γε καὶ ἀπὸ μανίας, ὥσπερ ἡ δύναμις, ἀνδρεία δὲ ἀπὸ φύσεως καὶ εὐτροφίας τῶν ψῦχων γίγνεται.

Pro. . . . When you asked me whether courageous men are bold, I admitted it: I was not asked whether bold men are courageous. Had you asked me this before, I should have said—'Not all.' And as to proving that courageous men are not bold, you have nowhere pointed out that I was wrong in my admission that they are. Next you show that such persons individually are bolder when they have knowledge, and bolder than others who lack it, and therewith you take courage and wisdom to be the same: proceeding in this manner you might even take strength to be wisdom. On this method you might begin by asking me whether the strong are powerful, and I should say 'Yes'; and then, whether those who know how to wrestle are more powerful than those who do not know how to wrestle, and whether individually they are more powerful when they have learnt than before learning, and I should say 'Yes'. And on my admitting these points it would be open to you to say, by the same token, that according to my admission, wisdom is strength. But neither there nor elsewhere do I admit that the powerful are strong, only that the strong are powerful; for I hold that power and strength are not the same, but that one of them, power, comes from knowledge, or from madness or rage, whereas strength comes from constitution and fit nurture of the body. So, in the other instance, boldness and courage are not the same, and therefor it results that the courageous are bold, but not that the bold are courageous; for boldness comes to a man from art, or from rage or madness, like power, whereas courage comes from constitution and fit nurture of the soul.

This third text shows in even clearer fashion than the second, so Bocheński writes, how difficult Plato found the question of whether 'All A is B' implies 'All B is A'. Just how difficult he found it is illustrated by the fact that 'to show the invalidity of the foregoing rule of conversion, he betakes himself to complicated extra-logical discussions—about bodily strength, for instance' (p. 42).

Here it seems somewhat more clear that Bocheński thinks that Plato has really perceived that A propositions are not convertible *simpliciter*, but because he found the whole question still so difficult and because, as a result of his difficulties, he brought in extra-logical points, we are apparently again intended to infer that Plato deserves no credit for this perception.

Now, since Plato's remarks on the subject of bodily strength seem to be regarded as a particular mark of his difficulties with the conversion rule, let us see whether these remarks really are extra-logical.

The general circumstances in which the passage 350Cff. occurs are as follows: in the earlier discussion with Protagoras (329B–334C) Socrates attempted (by, as I think, consciously fallacious means [see above, p.28]) to reduce three of the virtues to knowledge, i.e., justice, holiness, and temperance. In order to complete the refutation of Protagoras (who has maintained 330A that the virtues are distinct) it is necessary that the remaining virtue, courage, should be reduced to knowledge also. The reduction of courage is more difficult since as Protagoras says 349D, 'you will find many people extremely unjust, unholy, dissolute, and ignorant, and yet pre-eminently courageous'. It is thus postponed until after the discussion of Simonides' poem (339A–347A). When the argument is resumed at 349D, Socrates first attempts a very rapid completion of it at 349E–350C. After Protagoras rightly points out the false conversion which has been used, however, Socrates gives up this line of approach (very abruptly, in fact, at 351B) and embarks on a longer route to his objective, one which involves him in the discussion of pleasure which is the distinctive feature of the latter part of the dialogue.

At the moment we need to examine Socrates' argument 349E–350C (the attempted brief reduction of courage to knowledge) and to compare it with the remarks on bodily strength in Protagoras' speech of correction 350C–351B. We shall then see that what Protagoras has done is to give an argument precisely similar in structure to that of Socrates but one which is instantiated by different terms:

SOCRATES	PROTAGORAS
(*a*) All the courageous are bold.	(*a*) All the strong are powerful.
(*b*) those who know (e.g. how to dive into wells) are bolder than those who do not know, or All the knowing are bold.	(*b*) those who know (e.g. how to wrestle) are more powerful than those who do not know, or All the knowing are powerful.
(*c*) Some bold men are not courageous.	(*c*) (Some powerful men are not strong.)[50]
(*d*) All the bold are courageous.	(*d*) All the powerful are strong.
(*e*) All the knowing are courageous.	(*e*) All the knowing are strong.
(*f*) All knowledge is courage.	(*f*) All knowledge is strength.

The false conversion of (*a*) occurs at (*d*). (*d*) is then combined with (*b*) to produce, by a valid syllogism, the conclusion (*e*). (Whether (*f*) is the legitimate result of (*e*) is a debatable point, but not one which we need to consider here.) What Protagoras has done is, in fact, very similar to what Socrates has done at *Euthyphro* 12Aff. Euthyphro at first failed to understand what Socrates said about the holy and the right, but did understand when Socrates constructed a similar argument about reverence and fear. Protagoras shows, in the speech 350Cff., that he understands that Socrates' argument contains a false conversion, and he makes his understanding clear by constructing a similar argument with different terms. Bocheński has not quoted enough of the important passage; if we begin at 349E instead of 350C, we can see the parallelism I have just indicated. It then becomes clear that Plato's remarks about bodily strength are by no means extra-logical, unless of course one wants to say that all instantiation of logical laws is extra-logical. (In this case Bocheński's objection would apply to the *Gorgias* and *Euthyphro* passages as well.)

All in all, it is very difficult to avoid the conclusion that Fr. Bocheński's real quarrel with Plato is that this unenlightened

[50] This step does not occur in the text but has been added for the sake of filling out the scheme. I have also reworded some of the other steps (notably (*b*)) in order to show more clearly what I take to have been Plato's line of thought.

philosopher did not employ the notation of *Principia Mathematica* or something comparable. (In this respect he resembles Robinson, who thinks that Plato cannot have understood the importance of equivocation since he did not call it by a technical name [see above p. 7n.].) Plato's understanding of the conversion rule is too obvious for Bocheński to deny it outright, but he cannot, apparently, bring himself to congratulate Plato on the correct formulation of this rule since it is nowhere stated in the dialogues in technical form. Thus he confuses the issue with his talk about 'struggles' and 'difficulties', in this way contriving to give the impression that although some of these struggles may have met with success, Plato still remains a second- or third-rate logician. (The possibility that some of Plato's 'struggles' and 'difficulties' may have a connection with his dramatic technique has not been considered.)

In general I find myself in substantial agreement with what Mr. John Ackrill has written in his review of *Ancient Formal Logic* in *Mind*, LXII, 1953, p. 111:

> Fr. Bocheński's dismissal of Plato in two paragraphs is grotesquely cavalier. . . . The passages he refers to are quite inadequate to support such a sweeping stricture, even waiving the fact that in some of them Plato is exposing not committing fallacies. . . . Plato often uses arguments long and short which can readily be symbolised. He does not often state or discuss logical laws, but the laws in accordance with which he argues can be elicited.

In *Formale Logik*, as I have indicated, the stricture is not quite so sweeping, and the dismissal takes somewhat longer, but Fr. Bocheński still seems to be of the same mind in the matter of Plato's logical competence. I hope I have succeeded in demonstrating that the question is at least open to somewhat more lengthy debate.[51]

[51] In my symbolic notation I have followed Bocheński and have regularly omitted quantifiers.

General Index

Index of Passages

INDEX OF PASSAGES